GUIDED NOTEBOOK FOR MYMATHLAB®

BEGINNING & INTERMEDIATE ALGEBRA

Kirk Trigsted

University of Idaho

Randy Gallaher

Lewis & Clark Community College

Kevin Bodden

Lewis & Clark Community College

PEARSON

Boston Columbus Indianapolis New York San Francisco Upper Saddle River
Amsterdam Cape Town Dubai London Madrid Milan Munich Paris Montreal Toronto
Delhi Mexico City Sao Paulo Sydney Hong Kong Seoul Singapore Taipei Tokyo

Reproduced by Pearson from electronic files supplied by the author.

Publishing as Pearson, 75 Arlington Street, Boston, MA 02116.

ISBN-13: 978-0-321-73857-8
ISBN-10: 0-321-73857-8

1 2 3 4 5 6 EB 16 15 14 13 12

www.pearsonhighered.com

Table of Contents

Section R.1 Guided Notebook

Section R.1 Operations on Fractions

Read the list of "THINGS TO KNOW" and review any concepts you are unfamiliar with.

Section R.1 Objective 1: Write Fractions in Equivalent Form

Write down the definitions for the following terms.

Fraction

Numerator

Denominator

Proper fraction

Improper fraction

Mixed number

Example 1:

Study the solutions for example 1 on page R.1-3. Complete the chart on your own and check your answers by clicking on the link. If your answers are incorrect watch the video to find your error.

	Mixed Number or Whole Number	Improper Fraction
a.	$6\frac{7}{8}$	
b.		$\frac{137}{9}$
c.	$9\frac{7}{16}$	
d.		$\frac{78}{6}$

Section R.1

What are **equivalent fractions**?

How can equivalent fractions be obtained?

What is the **Property of Equivalent Fractions**?

Example 2:

Study the solution for example 2 part a on page R.1-4 and record the answer below. Complete part b on your own and check your answer by clicking on the link. If your answer is incorrect watch the video to find your error.

Find the equivalent fraction with the given numerator or denominator.

a. $\frac{9}{10}$, denominator 60 b. $\frac{24}{56}$, numerator 3

<u>**Section R.1 Objective 2:** Write Fractions in Simplest From</u>

When is a fraction in **simplest form**, or lowest terms?

Write the procedure for **Simplifying Fractions.**

Read and summarize the CAUTION statement on page R.1-5.

Example 3:
Study the solutions for example 3 parts a and b on page R.1-6 and record the answers below. Complete parts c and d on your own and check your answers by clicking on the link. If your answers are incorrect watch the video to find your error.

Write each fraction in simplest form.

a. $\frac{12}{14}$ b. $\frac{28}{45}$ c. $\frac{7}{21}$ d. $\frac{36}{54}$

Section R.1 Objective 3: Multiply and Divide Fractions

What is the procedure for **Multiplying Fractions**?

Example 4:
Study the solutions for example 4 parts a and b on pages R.1-7 and R.1-8 and record the answers below. Complete parts c and d on your own and check your answers by clicking on the link. If your answers are incorrect watch the video to find your error.

Multiply and simplify.

a. $\frac{5}{6} \cdot \frac{9}{25}$ b. $\frac{4}{45} \cdot \frac{15}{8}$ c. $18 \cdot \frac{5}{9}$ d. $\frac{5}{14} \cdot 6\frac{3}{10}$

Read the Note on page R.1-8. Rework parts a and b using this technique.

What are **reciprocals**, or multiplicative inverses?

How can you find the reciprocal of a nonzero fraction?

What is the procedure for **Dividing Fractions**?

Example 5:
Study the solution for example 5 part a on page R.1-9 and record the answer below. Complete parts b and c on your own and check your answers by clicking on the link. If your answers are incorrect watch the video to find your error.

Divide and simplify.

a. $\frac{2}{5} \div \frac{7}{3}$ b. $\frac{8}{10} \div 6$ c. $2\frac{3}{5} \div 3\frac{5}{7}$

Read and summarize the CAUTION statement on page R.1-10.

<u>**Section R.1 Objective 4:** Add and Subtract Fractions</u>

What are **like fractions**?

What is the procedure for **Adding and Subtracting Like Fractions**?

Example 6:
Study the solution for example 6 part a on page R.1-11 and record the answer below. Complete part b on your own showing all work below.

Add or subtract. Simplify if necessary.

a. $\frac{2}{15} + \frac{11}{15}$ b. $\frac{9}{10} - \frac{5}{10}$

What is a **least common denominator**?

According to the popup box on page R.1-12, what are the steps for finding the LCD?

1.

2.

3.

What are the **Steps for Adding or Subtracting Fractions**?

1.

2.

3.

4.

Example 7:
Study the solution for example 7 part a on page R.1-13 and record the answer below. Complete parts b and c on your own and check your answers by clicking on the link. If your answers are incorrect watch the video to find your error.

Add or subtract. Simplify if necessary.

a. $\frac{1}{12}+\frac{4}{9}$ b. $\frac{11}{30}-\frac{7}{24}$ c. $1\frac{5}{6}+4\frac{7}{9}$

Section R.2 Guided Notebook

Section R.2 Decimals and Percents

Read the list of "THINGS TO KNOW" and review any concepts you are unfamiliar with.

Section R.2 Objective 1: Add and Subtract Decimals

What is a **decimal**?

Recreate the **Place Value Chart** from page R.2-3.

What are the **Steps for Adding or Subtracting Decimals?**

1.

2.

3.

Example 1:
Study the solutions for example 1 parts a and b on page R.2-4 and record the answers below. Complete parts c and d on your own and check your answers by clicking on the link. If your answers are incorrect, watch the video to find your error.

Add or subtract.

a. $16.93 + 4.695$

b. $53.72 - 6.589$

c. $143 + 8.67$

d. $7.2 - 0.036$

Section R.2 Objective 2: Multiply and Divide Decimals

What are the **Steps for Multiplying Decimals?**

1.

2.

What are the **Steps for Dividing Decimals**?

1.

2.

3.

Example 2:
Study the solutions for example 2 parts a and b on page R.2-6 and record the answers below. Complete parts c and d on your own and check your answers by clicking on the link. If your answers are incorrect, watch the video to find your error.

Multiply or divide.

a. $2.5 \cdot 3.7$ b. $8.925 \div 7.5$ c. $1.49 \cdot 0.04$ d. $7.5 \div 0.33$

Section R.2 Objective 3: Convert Between Decimals and Fractions

Launch the popup box on page R.2-8 and record the steps to write a decimal as a fraction.

1.

2.

3.

Example 3:
Study the solution for example 3 part a on page R.2-8 and record the answer below. Complete parts b and c on your own and check your answers by clicking on the link. If your answers are incorrect, watch the video to find your error.

Write each decimal as a fraction. Simplify if necessary.

a. 0.23 b. 0.125 c. 8.6

What is the procedure to write a fraction as a decimal?

The quotient will either be a ______________________ or ________________ decimal.

Example 4:
Study the solution for example 4 part a on page R.2-9 and record the answers below. Complete parts b and c on your own and check your answers by clicking on the link. If your answers are incorrect, watch the video to find your error.

Write each fraction as a decimal.

a. $\frac{9}{11}$ b. $\frac{7}{20}$ c. $\frac{19}{8}$

Read and summarize the CAUTION statement on page R.2-10. Show an example.

Section R.2 Objective 4: Convert Between Percents and Decimals or Fractions

What does the word **percent** mean?

What is the **percent symbol**?

Multiplying or dividing by 100% is equivalent to what?

What is the procedure for **Writing a Percent as a Decimal or Fraction**?

According to the **Note** this is the same as what?

What is the procedure for **Writing a Decimal or Fraction as a Percent**?

According to the **Note** this is the same as what?

Read and summarize the CAUTION statement on page R.2-11

Example 5:
Study the solutions for example 5 parts a and b on page R.2-12 and record the answers in the chart below. Complete parts c and d on your own and check your answers by clicking on the link. If your answers are incorrect watch the video to find your error.

Complete the chart with appropriate percents, decimals, and fractions.

	Percent	**Decimal**	**Fraction**
a.	28%		
b.			$\frac{17}{40}$
c.		0.875	
d.			$\frac{13}{12}$

Section 1.1 Guided Notebook

Section 1.1 The Real Number System

Read the list of "THINGS TO KNOW" and review any concepts you are unfamiliar with.

Section 1.1 Objective 1: Classify Real Numbers

Write down the definitions for each of the following terms.

Set

Element

Empty set; null set

Null symbol

Finite sets

Infinite sets

Natural Numbers

Whole numbers

Integers

Rational numbers

Irrational numbers

Real numbers

Read and summarize the CAUTION statements on pages 1.1-3 and 1.1-7.

1.1-3

1.1-7

Recreate figure 1 on page 1.1-8, showing the relationships involving the set of real numbers.

Example 1:
Study the solutions for example 1 parts a–c on page 1.1-9 and record the answers below. Complete parts d–h on your own and check your answers by clicking on the link. If your answers are incorrect, watch the video to find your error.

Classify each **real number** as a **natural number**, **whole number**, **integer**, **rational number** and/or **irrational number**. Each number may belong to more than one set.

a. 8

b. -4.8

c. $\sqrt{10}$

d. -7

e. $-\frac{4}{7}$

f. $\sqrt{25}$

g. 0

h. $3.\overline{45}$

Read and summarize the CAUTION Statement on page 1.1-10.

<u>**Section 1.1 Objective 2:** Plot Real Numbers on a Number Line</u>

What is a **real number line**?

What are the point, and the real number corresponding to the point, called?

What is the **origin**?

Is 0 positive or negative?

What are **consecutive integers**?

Section 1.1

What does it mean to **plot** or **graph**?

Draw the real number line labeling the following: Negative Real Numbers, Positive Real Numbers, Zero and the Origin.

Example 2:
Work through example 2 on page 1.1-12 and record the answer below.

Plot the following **set** of numbers on the **number line**.

$\{-3, -\frac{3}{2}, 1, 2.25\}$

Sections 1.1 Objective 3: Find the Opposite of a Real Number

What are **opposites**?

What is used to represent "the opposite of"?

What is the **Double-Negative Rule**?

What is the procedure for **Finding the Opposite of a Real Number**?

Example 3:
Work through example 3 on page 1.1-15 and record the answers below.

Find the opposite of each real number.

a. 22 b. $-\frac{4}{5}$ c. 6.4 d. 0

Section 1.1 Objective 4: Find the Absolute Value of a Real Number

What is **unit distance**?

What is the definition of **absolute value**?

Read the CAUTION statement on page 1.1-16.

Example 4:
Study the solutions for example 4 parts a and b on page 1.1-17 and record the answers below. Complete parts c–e on your own and check your answers by clicking on the link. If your answers are incorrect, watch the video to find your error.

Find each absolute value.

a. $|3|$ b. $|5|$ c. $|-1.5|$ d. $\left|-\frac{7}{2}\right|$ e. $|0|$

<u>**Section 1.1 Objective 5:** Use Inequality Symbols to Order Real Numbers</u>

What does it mean to **order** two real numbers?

How do you find the order of any two real numbers?

What is the **equal sign**?

What are **inequality symbols**? Identify them as “less than” or “greater than”.

Example 5:
Work through example 5 on page 1.1-20 and record the answers below.

Fill in the blank with the correct symbol, <, >, or = to make a true statement.

a. 0_____3 b. –3.7_____–1.5 c. $-\frac{5}{4}$_____-1.25 d. $\frac{4}{5}$_____$\frac{5}{9}$

What are **non-strict inequalities**?

What are **strict inequalities**?

Example 6:
Study the solutions for example 6 parts a and b on pages 1.1-22 and 1.1-23 and record the answers below. Complete parts c–e on your own and check your answers by clicking on the link. If your answers are incorrect, watch the video to find your error.

Determine if each statement is true or false.

a. $\frac{7}{10} \le 0.7$ b. $-8 \ge 4$ c. $-2 \ge -4$ d. $\frac{7}{3} \le 1.\overline{3}$ e. $-\frac{9}{4} \ne -2.75$

Section 1.1 Objective 6: Translate Word Statements Involving Inequalities

Complete the following table as found on page 1.1-24.

Inequality	Key Words	Word Statement	Mathematical Expressions
		Six is fewer than eight	
$>$			
	is less than or equal to, at most, no more than		
			$15 \geq 13$; $8 \geq 3$
		12 is not equal to 15	

Example 7:
Work through example 7 on page 1.1-25 and record the answers below. Check your answers by clicking on the link.

Write a mathematical expression for each work phrase.

a. Fourteen is greater than ten.

b. Twenty-four is no more than thirty.

c. Nine is not equal to eighteen.

Example 8:
Study the solution for example 8 part a on page 1.1-26 and record the answer below. Complete parts b and c on your own and check your answers by clicking on the link. If your answers are incorrect, watch the video to find your error.

Use real numbers and write an inequality that represents the given comparison.

a. Orchid Island Gourmet Orange Juice sells for \$6, which is more than Florida's Natural Premium Orange Juice that sells for \$3.

b. In 2008 there were 4983 identity thefts reported in Colorado, which is different than the 4433 reported in Missouri during the same year.

c. On May 20, 2010, the Dow Jones Industrial Average closed at 10,068.01 which was less than on May 19, 2010, when it closed at 10,444.37.

Section 1.2 Guided Notebook

Section 1.2 Adding and Subtracting Real Numbers

Read the list of "THINGS TO KNOW" and review any concepts you are unfamiliar with.

Section 1.2 Objective 1: Add Two Real Numbers with the Same Sign

What is result of adding numbers?

What are the numbers being added called?

What are the three steps for **Adding Two Real Numbers with the Same Sign**?

1.

2.

3.

Example 1:
Work through example 1 page 1.2-3 and record the answers below. View the popups on page 1.2-4 to see the visualization of examples 1a and 1b.

Add

a. $2 + 5$

b. $-4 + (-3)$

View and take notes on the example on page 1.2-4, using money to think about adding real numbers.

Example 2:
Study the solutions for example 2 parts a and b on page 1.2-5 and record the answers below. Complete parts c and d on your own and check your answers by clicking on the link. If your answers are incorrect, watch the video to find your error.

Add.

a. $-3.65 + (-7.45)$

b. $\frac{4}{5} + \frac{13}{5}$

c. $-\frac{3}{5} + \left(-\frac{7}{2}\right)$

d. $-3\frac{1}{3} + \left(-5\frac{1}{4}\right)$

Section 1.2 Objective 2: Add Two Real Numbers with Different Signs

What are the three steps for **Adding Two Real Numbers with Different Signs?**

1.

2.

3.

Example 3:
Work through example 3 on page 1.2-7 and record the answers below. View the popups on page 1.2-9 to see the visualization of example 3a and 3b on a number line.

Add.

a. $7 + (-4)$ b. $-6 + 4$ c. $6 + (-6)$

What are **additive inverses**?

What is the rule for **Adding a Real Number and Its Opposite**?

What is the procedure for **Adding Two Real Numbers**?

1.

2.

Example 4:
Study the solutions for example 4 parts a and b on pages 1.2-10 and 1.2-11 and record the answers below. Complete parts c and d on your own and check your answers by clicking on the link. If your answers are incorrect, watch the video to find your error.

Add.

a. $-12 + (-9)$ b. $7 + (-18)$ c. $\frac{4}{3} + \frac{5}{6}$ d. $-5.7 + 12.3$

Section 1.2 Objective 3: Subtract Real Numbers

What is the **difference**?

What is the **subtrahend**?

What is the **minuend**?

What is the definition of **Subtracting Two Real Numbers**?

Example 5:

Study the solutions for example 5 parts a–d on pages 1.2-13 and 1.2-14 and record the answers below. Complete parts e–g on your own and check your answers by clicking on the link. If your answers are incorrect, watch the video to find your error.

Subtract.

a. $15 - 6$ b. $9 - 17$ c. $-4.5 - 3.2$ d. $\frac{7}{3} - \frac{2}{5}$

e. $-4.9 - (-2.5)$ f. $7\frac{3}{4} - \left(-2\frac{1}{5}\right)$ g. $4 - (-4)$

Section 1.2 Objective 4: Translate Word Statements Involving Addition or Subtraction

Complete the table below found on page 1.2-15.

Key Word	Word Phrase	Mathematical Expression
Sum		
		$4 + 7$
	3 *added to* 8	
		$-4 + 6$

Example 6:

Work through example 6 on page 1.2-15 and record the answers below.

Write a mathematical expression for each word phrase.

a. Five more than –8

b. 8.4 increased by 0.17

c. The sum of –4 and –10

d. 15 added to –30

Complete the table below found on page 1.2-18.

Key Word	**Word Phrase**	**Mathematical Expression**
		9 – 7
Decreased by		
	6 *subtracted from* 3	
		25 – 10

Read and summarize the CAUTION statement on page 1.2-18.

Example 7:

Study the solutions for example 7 parts a–c on pages 1.2-19 and 1.2-20 and record the answers below. Complete parts d and e on your own and check your answers by clicking on the link. If your answers are incorrect, watch the video to find your error. Be sure to note the key words and the order of the subtraction.

Write a **mathematical expression** for each word phrase.

a. Fifteen subtracted from 22.

b. The **difference** of 7 and 12.

c. Eight decreased by 11.

d. 20 less than the difference of 4 and 9.

e. The **sum** of 8 and 13, decreased by 5.

Section 1.2 Objective 5: Solve Applications Involving Addition or Subtraction of Real Numbers

Example 8:
Work through example 8 on page 1.2-22 and record the answer below.

On May 20, 2010 the Dow Jones Industrial Average lost 376.36 points. The next day it gained 125.38 points. What was the total result for the two days?

Example 9:
Work through Example 9 on page 1.2-23 and record the answer below. Check your answer by clicking on the link. If your answer is incorrect, watch the video to find your error.

The record high temperature in Alaska was $100^{\circ}F$ recorded in 1915 at Fort Yukon. The record low was $-80^{\circ}F$ recorded in 1971 at Prospect Creek Camp. What is the difference between these record high and low temperatures?

Section 1.3 Guided Notebook

Section 1.3 Multiplying and Dividing Real Numbers

Read the list of "THINGS TO KNOW" and review any concepts you are unfamiliar with.

Section 1.3 Objective 1: Multiply Real Numbers

Multiplication is simply repeated _______________.

What is the **product**?

What are **factors**?

What does the sign of the product depend on?

Example 1:
Work through example 1 on page 1.3-4 and record the answers below.

Multiply.

a. $5\cdot 13$ b. $6(-7)$ c. 0×15 d. $(-4)(-12)$

Example 2:
Study the solution for example 2 part a on page 1.3-5 and record the answer below. Complete parts b–d on your own and check your answers by clicking on the link. If your answers are incorrect, watch the video to find your error.

Multiply.

a. $\left(-\frac{3}{4}\right)\left(-\frac{7}{9}\right)$ b. $5\cdot\frac{3}{10}$ c. $\frac{3}{8}\times 0$ d. $\left(-\frac{2}{3}\right)\left(\frac{6}{14}\right)$

Example 3:
Study the solutions for example 3 parts a and b on page 1.3-6 and record the answers below. Complete parts c and d on your own and check your answers by clicking on your link. If your answers are incorrect, watch the video to find your error.

Multiply.

a. (1.4)(–3.5) b. (8.32)(0) c. $-\frac{3}{5}\times 6\frac{1}{3}$ d. (4)(5.8)

Section 1.3 Objective 2: Divide Real Numbers

What is a **quotient**?

What is the **dividend**?

What is the **divisor**?

What is the **reciprocal** or **multiplicative inverse**?

What is the definition of **reciprocals**?

Click on the link for the first CAUTION statement on page 1.3-8, and show why 0 does not have a reciprocal.

Summarize the second CAUTION statement on page 1.3-8.

What is the definition of **Division of Two Real Numbers**?

Example 4:
Work through example 4 on page 1.3-9 and record the answers below.

Divide.

a. $\dfrac{-60}{4}$

b. $\dfrac{3}{25} \div \left(-\dfrac{9}{20}\right)$

Example 5:
Study the solutions for example 5 parts a and b on page 1.3-11 and record the answers below. Complete parts c and d on your own and check your answers by clicking on the link. If your answers are incorrect, watch the video to find your error.

Divide.

a. $(-8) \div (-36)$

b. $\dfrac{0}{5}$

c. $\dfrac{12}{35} \div \left(-\dfrac{27}{14}\right)$

d. $15 \div \left(\dfrac{3}{4}\right)$

Example 6:
Study the solutions for example 6 parts a and b on page 1.3-12 and record the answers below. Complete parts c and d on your own and check your answers by clicking on the link. If your answers are incorrect, watch the video to find your error.

Divide.

a. $\dfrac{48.6}{-3}$

b. $-7\dfrac{2}{5} \div (-3)$

c. $\dfrac{-59.4}{4.5}$

d. $6\dfrac{5}{8} \div 2\dfrac{1}{4}$

According to the CAUTION statement on page 1.3-13, what are the three ways to write a negative quotient?

Section 1.3 Objective 3: Translate Word Statements Involving Multiplication or Division

Complete the table below that is found on page 1.3-14.

Key Word	Word Phrase	Mathematical Expression
Product		
	–7 *times* –9	
	One-third *of* 27	
		0.15(200)
Twice		

What are four symbols that can be used to indicate multiplication?

Example 7:
Study the solutions for example 7 parts a–d on page 1.3-15 and record the answers below. Complete parts e and f on your own and check your answers by clicking on the link. If your answers are incorrect, watch the video to find your error.

Write a mathematical expression for each word phrase.

a. The product of 3 and –6.

b. 30% of 50

c. Three times the sum of 10 and 4

d. Three-fourths of 20, increased by 7

e. The difference of 2 and the product of 8 and 15

f. 3 increased by 15, times 4

Rework part d as "three-fourths of 20 increased by 7." To see the difference, click on the link on page 1.3-18.

Section 1.3

Complete the table below from page 1.3-18.

Key Word	Word Phrase	Mathematical Expression
Quotient		
	24 *divided by* -3	
Per		
		$\frac{4}{9}$

Example 8:
Study the solutions for example 8 parts a and b on page 1.3-19 and record the answers below. Complete parts c and d on your own and check your answers by clicking on the link. If your answers are incorrect, watch the video to find your error.

Write a mathematical expression for each word phrase.

a. The ratio of 10 to 35

b. 60 divided by the sum of 3 and 7

c. The quotient of 20 and 4

d. The difference of 12 and 7, divided by the difference of 8 and -3

Rework part 8d as "the difference of 12 and 7 divided by the difference of 8 and -3." Click on the link on page 1.3-20 to see the difference.

Section 1.3 Objective 4: Solve Applications Involving Multiplication or Division

Example 9:
Work through example 9 on page 1.3-21 and record your answer below.

For their 8 home games in the 2009 regular season, the Denver Broncos had a total attendance of 600,928. Assuming equal attendance at all home games, how many people attended each home game?

Example 10:
Work through example 10 on page 1.3-22 and record the answer below.

The amount of acid in a solution can be found by multiplying the volume of solution by the percent of the solution (written in decimal form). How much acid is in 20 liters of a 3% solution?

Section 1.4 Guided Notebook

Section 1.4 Exponents and Order of Operations

Read the list of "THINGS TO KNOW" and review any concepts you are unfamiliar with.

Section 1.4 Objective 1: Evaluate Exponential Expressions

Write down an exponential expression. Label and explain the meaning of each of the following in your expression: **exponent, power,** and **base**.

What is the definition of an **exponential expression**?

When the exponent is 1, the number is raised to the ________ power. In this case, the real number raised to the _______ power is equal to ____________.

If no exponent is written it is assumed to be _____.

Example 1:
Work through example 1 on page 1.4-4 and record the answers below.

Evaluate each exponential expression.

a. 4^3 b. $\left(\frac{2}{3}\right)^4$ c. $(0.3)^2$

When is a negative sign part of the base? Write an example showing each situation.

Example 2:
Study the solutions for example 2 parts a and b on page 1.4-6 and record the answers below. Complete part c on your own and check your answer by clicking on the link. If your answer is incorrect, watch the video to find your error.

Evaluate each exponential expression.

a. $(-5)^3$ b. -4^2 c. $(-2)^4$

View the link on page 1.4-6 to see how an **exponent** is used to determine the sign when the base of an exponential expression is negative.

Section 1.4 Objective 2: Use the Order of Operations to Evaluate Numeric Expressions

What is the **order of operations**?

View and summarize the **tip** in the link on page 1.4-7.

Example 3:
Work through example 3 on page 1.4-8 and record the answers below.

Simplify each expression.

a. $10-4^2$

b. $12\div4+8$

Take notes on the illustration in each of the two links on page 1.4-9.

Example 4:
Work through example 4 on page 1.4-9 and record the answers below. Check your answers by clicking on the link. If your answers are incorrect, watch the video to find your error.

Simplify each expression.

a. $15-3+6-8+7$

b. $3\cdot15\div5\cdot6\div2$

Example 5:
Work through example 5 on page 1.4-9 and record the answers below.

Simplify each expression.

a. $5+(4-2)^2-3^2$ b. $[5-9]^2+12\div 4$

Example 6:
Work through example 6 on page 1.4-11 and record the answers below. Check your answers by clicking on the link. If your answers are incorrect, watch the video to find your error.

Simplify each expression.

a. $(-5+8)\cdot 3$ b. $(10-4)^2$ c. $12\div(4+8)$

Write down five **grouping symbols** and their names.

View the **calculator example** shown the CAUTION statement on page 1.4-11, discuss the importance of grouping symbols when using a calculator.

In fractions, simplify the ________________ and ________________ before ___________.

In absolute values, first simplify expressions _____________ absolute value symbols, before ________________ the absolute value.

Read and summarize the CAUTION statement on page 1.4-12.

Example 7:
Study the solution for example 7 part a on page 1.4-13 and record the answer below. Complete part b on your own and check your answer by clicking on the link. If your answer is incorrect, watch the video to find your error.

Simplify each expression.

a. $\dfrac{-2(3)+6^2}{(-4)^2-1}$

b. $|7^2-5(3)|\div 2+8$

What are **nested grouping symbols**?

Example 8:
Study the solution for example 8 part a on page 1.4-14 and record the answer below. Complete part b on your own and check your answer by clicking on the link. If your answer is incorrect, watch the video to find your error.

Simplify each expression.

a. $\left[2^3-3(5-7)^2\right]\div 6-9$

b. $\dfrac{|-5^2+2^3|-10}{4^2-6\cdot 5}$

Example 9:
Work through example 9 on page 1.4-15 and record the answers below. Check your answers by clicking on the link. If your answers are incorrect, watch the video to find your error.

Simplify each expression.

a. $\dfrac{3}{10}\cdot\dfrac{5}{2}-\dfrac{1}{2}$

b. $36\div\dfrac{8}{3^2-5}+(-2)^3$

c. $\dfrac{\left|\dfrac{1}{3}-\dfrac{3}{5}\right|}{4}\div\dfrac{1}{2}-1$

Section 1.5 Guided Notebook

Section 1.5 Variables and Properties of Real Numbers

Read the list of "THINGS TO KNOW" and review any concepts you are unfamiliar with.

Section 1.5 Objective 1: Evaluate Algebraic Expressions

Write down the definitions for the following terms.

Variable

Constant

Algebraic expression

Give four examples of algebraic expressions and their meanings.

What operation is performed when a constant appears next to a variable?

Describe the process to **Evaluate Algebraic Expressions**.

Example 1:
Work through example 1 on page 1.5-5 and record your answers below.

Evaluate each algebraic expression for the given values of the variables.

a. $3x + 7$ for $x = 5$

b. $a^2 - 3$ for $a = -4$

Read and summarize the CAUTION statement on page 1.5-5.

Example 2:
Study the solution for example 2 part a on page 1.5-6 and record the answer below. Complete part b on your own and check your answer by clicking on the link. If your answer is incorrect, watch the video to find your error.

Evaluate each algebraic expression for the given values of the variables.

a. $5x^2 + 9$ for $x = 7$

b. $-2(m + 3) - 5$ for $m = 8$

Example 3:
Study the solution for example 3 part a on page 1.5-7 and record the answer below. Complete part b on your own and check your answer by clicking on the link. If your answer is incorrect, watch the video to find your error.

Evaluate each algebraic expression for the given values of the variables.

a. $\frac{x^2+6}{5x-2}$ for $x = 2$

b. $|3y-4|+7y-1$ for $y = -3$

Example 4:
Study the solution for example 4 part a on page 1.5-8 and record the answer below. Complete part b on your own and check your answer by clicking on the link. If your answer is incorrect, watch the video to find your error.

Evaluate each algebraic expression for the given values of the variables.

a. $12a + 7b$ for $a = -4$ and $b = 12$

b. $x^2 - 2xy + 3y^2$ for $x = 3$ and $y = -1$

Section 1.5 Objective 2: Use the Commutative and Associative Properties

Describe the **Commutative Property of Addition** and show an example.

Describe the **Commutative Property of Multiplication** and show an example.

According to the CAUTION statement on page 1.5-10, what operations **do not** have commutative properties? Why not?

Example 5:
Work through example 5 on page 1.5-11 and record the answers below.

Use the given property to rewrite each statement. Do not simplify.

a. Commutative property of multiplication: –2 (6) = ____________

b. Commutative property of addition: 5.03 + 9.2 = ____________

Describe the **Associative Property of Addition** and show an example.

Describe the **Associative Property of Multiplication** and show an example.

According to the CAUTION statement on page 1.5-12, what operations **do not** have associative properties? Why not?

Example 6:
Work through example 6 on page 1.5-13 and record the answers below.

Use the given property to rewrite each statement. Do not simplify.

a. Associative property of addition: $\left(\frac{2}{3}+\frac{1}{6}\right)+\frac{5}{6}=$ __________________

b. Associative property of multiplication: $5\cdot(2\cdot 13)=$ _______________

Example 7:
Work through example 7 on page 1.5-14 and record the answers below.

Use the commutative and associative properties to simplify each expression.

a. $(3+x)+7$

b. $(8y)\left(\frac{1}{2}\right)$

Section 1.5 Objective 3: Use the Distributive Property

Describe the **Distributive Property** and show an example. What does it state?

Show two ways to write the Distributive Property.

Does the Distributive Property apply to subtraction? If yes, show an example.

According to the pop up on page 1.5-17, why does the Distributive Property extend to more than two terms?

Example 8:
Work through example 8 on page 1.5-17 and record the answers below.

Use the distributive property to remove parentheses, and write the product as a sum. Simplify if possible.

a. $9(x + 2)$ b. $(7x-5)\cdot 3$

How do you find the **Opposite of an Expression**?

Example 9:
Study the solution for example 9 part a on page 1.5-19 and record the answer below. Complete parts b and c on your own and check your answers by clicking on the link. If your answers are incorrect, watch the video to find your error.

Use the distributive property to remove parentheses, and write the product as a sum. Simplify if possible.

a. $2(4y + 3z - 5)$ b. $-6(3y - 8)$ c. $-(2a - 7b + 8)$

Example 10:
Work through example 10 on page 1.5-20 and record the answers below.

Use the distributive property to write each sum as a product.

a. $9\cdot x+9\cdot 4$ b. $4x + 4y$

Section 1.5 Objective 4: Use the Identity and Inverse Properties

Describe the **Identity Property of Addition** and show an example.

Describe the **Identity Property of Multiplication** and show an example.

Describe the **Inverse Property of Addition** and show an example.

Describe the **Inverse Property of Multiplication** and show an example.

Example 11:
Work through example 11 on page 1.5-23 and record the answers below. Check your answers by clicking on the link. If your answers are incorrect, watch the video to find your error.

Identify the property of real numbers illustrated in each statement.

a. $-4 \cdot 1 = -4$

b. $(-5+5)+x=0+x$

c. $0+y=y$

d. $\frac{1}{2} \cdot 2x = x$

Section 1.6 Guided Notebook

Section 1.6 Simplify Algebraic Expressions

Read the list of "THINGS TO KNOW" and review any concepts you are unfamiliar with.

Section 1.6 Objective 1: Identify Terms, Coefficients, and Like Terms of an Algebraic Expression

Write down the definitions for the following.

Terms

Variable terms

Constant terms

Coefficient

Read and summarize the CAUTION statement on page 1.6-4.

Example 1:
Study the solutions example 1 parts a and b on pages 1.6-4 and 1.6-5 and record the answers below. Complete part c on your own and check your answer by clicking on the link. If your answer is incorrect, watch the video to find your error.

Determine the number of terms in each expression and list the coefficients for each term.

a. $3x^2 + 7x - 3$

b. $4x^3 - \frac{3}{2}x^2 + x - 1$

c $3x^2 - 2.3x + x - \frac{3}{4}$

What are **like terms**? Show a pair of like terms and a pair of unlike terms.

Are constants like terms?

Example 2:
Study the solution for example 2 part a on page 1.6-6 and record the answer below. Complete part b on your own and check your answer by clicking on the link. If your answer is incorrect, watch the video to find your error.

Identify the like terms in each algebraic expression.

a. $5x^2 + 3x - 6 + 4x^2 - 7x + 10$

b. $3.5a^2 + 2.1ab + 6.9b^2 - ab + 8a^2$

Section 1.6 Objective 2: Simplify Algebraic Expressions

How are expressions simplified?

Example 3:
Study the solutions for example 3 parts a and b on pages 1.6-7 and 1.6-8 and record the answers below. Complete parts c–e on your own and check your answers by clicking on the link. If your answers are incorrect, watch the video to find your error.

Simplify each algebraic expression by combining like terms.

a. $5x - 2x$

b. $6x^2 - 12x - 3x^2 + 4x$

c. $3z - 2z^2 + 7z^2$

d. $6x^2 + 2x + 4x + 3$

e. $-3x + 5 - y + x - 8$

When is an algebraic expression **simplified**?

What are the steps for **Simplifying an Algebraic Expression**?

1.

2.

Section 1.6

Study the solutions for example 4 parts a and b on page 1.6-9 and record the answers below. Complete parts c and d on your own and check your answers by clicking on the link. If your answers are incorrect, watch the video to find your error.

Simplify each algebraic expression.

a. $3(x-4)+2$

b. $8(x+6)+7x$

c. $5(x-6)-3(x-7)$

d. $2(5z+1)-(3z-2)$

Section 1.6 Objective 3: Write Word Statements as Algebraic Expressions

Write down six key words that indicate addition.

Write down six key words that indicate subtraction.

Write down seven key words that indicate multiplication.

Write down five key words that indicate division.

What can we use to represent unknown quantities within a verbal description?

Example 5:
Study the solutions for example 5 parts a–c on pages 1.6-12 and 1.6-13 and record the answers below. Complete parts d–f on your own and check your answers by clicking on the link. If your answers are incorrect, watch the video to find your error.

Write each word statement as an **algebraic expression**. Use x to represent the unknown number.

a. Twenty decreased by a number

b. The product of sixteen and a number

c. Five more than twice a number

d. Three-fourths of the square of a number

e. The quotient of 12 and a number, increased by the number

f. The sum of a number and 4, divided by the difference of the number and 9

Example 6:
Study the solutions to example 6 parts a and b on page 1.6-14 and record the answers below. Complete part c on your own and check your answer by clicking on the link. If your answer is incorrect, watch the video to find your error.

a. The longest side of a triangle is four units longer than five times the length of the shortest side. Express the length of the longest side in terms of the shortest side, a.

b. Michelle invests d dollars in one account and \$6750 less than this amount in the second account. Express the amount she invests in the second account in terms of d.

c. The state of Texas has 10 fewer institutes of higher education than twice the number in Virginia. If we let n = the number of institutes in Virginia, express the number in Texas, in terms of n.

Section 1.6 Objective 4: Solve Applied Problems Involving Algebraic Expressions

Example 7:
Work through example 7 on page 1.6-15 and record the answers below.

The perimeter of a rectangle is the sum of the lengths of the sides of the rectangle. Use the following rectangle to answer the questions.

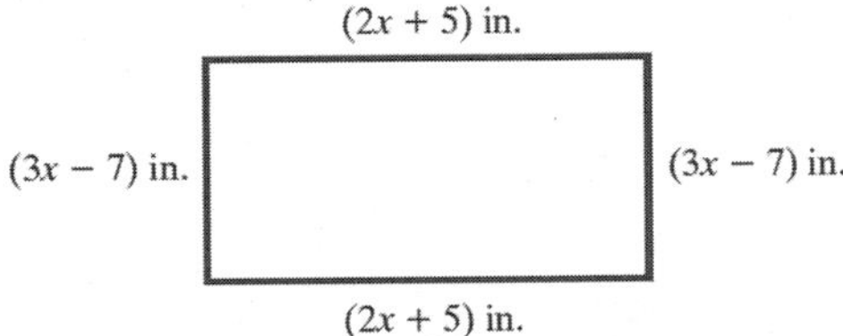

a. Write a simplified algebraic expression that represents the perimeter of the rectangle.

b. Use your result from part (a) to find the perimeter if $x = 7$.

Example 8:
Work through example 8 on page 1.6-17 and record the answers below. Check your answers by clicking on the link. If your answers are incorrect, watch the video to find your error.

Based on data from the National Fire Protection Association, the number of residential property fires, in thousands, is given by $111x^2 - 1366x + 6959$, where x = the number of years after 2000. The number of vehicle fires, in thousands, is given by $-52.5x + 775$.

a. Write a simplified algebraic expression for the difference between the number of residential property fires and the number of vehicle fires.

b. Use your result from part (a) to estimate the difference in 2010.

Section 2.1 Guided Notebook

Section 2.1 The Addition and Multiplication Properties of Equality

Read the list of "THINGS TO KNOW" and review any concepts you are unfamiliar with.

Section 2.1 Objective 1: Identify Linear Equations in One Variable
Write down the definitions for the following terms.
Equation

Algebraic equation

What is the difference between an algebraic expression and an algebraic equation?

Write down the algebraic equations from the interactive video on page 2.1-3.

What is the definition of a **linear equation in one variable?**

Can the variable occur multiple times in an **equation in one variable**?

Write down the definitions for the following terms.
First-degree equation

Nonlinear equations

According to the popup on page 2.1-4, what are three of the five examples of nonlinear equations? Why are they nonlinear?

Watch the **interactive video** on page 2.1-5 and determine which types of equations are linear and which are nonlinear. Copy down any linear equations from the video.

Example 1:
Work through example 1 on page 2.1-5 and record the answers below.

Determine if each is a linear equation in one variable. If not, state why.

a. $4x + 3 - 2x$

b. $4x + 2 = 3x - 1$

c. $x^2 + 3x = 5$

d. $2x + 3y = 6$

Section 2.1 Objective 2: Determine If a Given Value Is a Solution to an Equation

Show the steps to show the truth of each of the following equations. View the popup to check.

$-3+(-2)^2 = -7+1$

$|10-26|+4 = 3^2+11$

What is the definition of **solve**?

What is the definition of a **solution**?

How do you determine if a given value is a solution to an equation?

Read and summarize the CAUTION statement on page 2.1-7.

Example 2:
Study the solutions for example 2 parts a and b on page 2.1-7. Complete parts c and d on your own and check your answers by clicking on the link. If your answers are incorrect, watch the video to find your error.

Determine if the given value is a solution to the equation.

c. $|a-6|-1 = 9+a^2;\ a=-2$

d. $\frac{3}{5}w - \frac{1}{2} = -\frac{3}{10}w;\ w = \frac{5}{9}$

Sections 2.1 Objective 3: Solve Linear Equations Using the Addition Property of Equality

What is a **solution set**?

What are **equivalent equations**?

What are the two forms of an **isolated variable**?

Describe the concept of the balance scale shown in figure 1 on page 2.1-9.

What is the **Addition Property of Equality**?

Does this property hold true for subtraction? Why or why not?

Example 3:
Work through example 3 on page 2.1-11 and record the answers below. See how to *check* your answers by clicking on the link.

Solve.

a. $x - 5 = 3$

b. $y + \frac{2}{3} = \frac{1}{5}$

Example 4:
Work through example 4 part a on page 2.1-13 and record the answer below. Complete part b on your own and check your answer by clicking on the link.

Solve.

a. $3z - 4.5 = 4z$

b. $12 - 7x = -6x$

Example 5:
Work through example 5 on page 2.1-14 and record the answers below. Check your answers by clicking on the link. If your answers are incorrect, watch the video to find your error.

a. $5x - 3 = 6x + 2$

b. $-5w + 27 = 13 - 4w$

Section 2.1 Objective 4: Solve Linear Equations Using the Multiplication Property of Equality

What is the **Multiplication Property of Equality**?

Does this hold true for division? Why or why not?

According to the CAUTION statement on page 2.1-15, why is the use of zero not allowed?

Example 6:
Work through example 6 on page 2.1-16 and record the answers below. See how to *check* your answers by clicking on the link.

Solve.

a. $\frac{x}{5} = -3$

b. $-4x = -32$

Example 7:
Work through example 7 on page 2.1-18 and record the answers below. Check your answers by clicking on the link. If your answers are incorrect, watch the video to find your error.

Solve.

a. $\frac{4}{3}x = 52$

b. $2.2x = 6.93$

Section 2.1 Objective 5: Solve Linear Equations Using Both Properties of Equality

The ____________________ property is used first for what purpose?

The ____________________ property is then used for what purpose?

Example 8:
Work through example 8 on page 2.1-19 and record the answers below. See how to *check* your answers by clicking on the link.

Use the properties of equality to solve each equation.

a. $\frac{2}{5}x-4=4$

b. $7n=3n-8$

Example 9:
Study the solution to example 9 part a on page 2.1-21 and record the answer below. Complete part b on your own and check your answer by clicking on the link. If your answer is incorrect, view the video to find your error.

Use the properties of equality to solve each equation.

a. $7y+4=2y-6$

b. $2x-14.5=0.5x+50$

Section 2.2 Guided Notebook

Section 2.2 Solving Linear Equations in One Variable

Read the list of "THINGS TO KNOW" and review any concepts you are unfamiliar with.

Section 2.2 Objective 1: Solve Linear Equations Containing Non-Simplified Expressions
What should be done with non-simplified expressions before using the properties of equality?

Example 1:
Work through example 1 on page 2.2-3 and record the answer below. Check your answer by clicking on the link.

Solve: $4x + 7 - 2x = 5 - 3x - 3$

What should be done if the equation contains grouping symbols?

Example 2:
Work through example 2 on page 2.2-4 and record the answer below. Check your answer by clicking on the link. If your answer is incorrect, watch the video to find your error.

Solve: $7 - 2(4z - 3) = 3z + 1$

Example 3:
Work through example 3 on page 2.2-5 and record the answer below. Check your answer by clicking on the link. If your answer is incorrect, watch the video to find your error.

Solve: $2(3x - 1) - 5x = 3 - (3x + 1)$

Section 2.2 Objective 2: Solve Linear Equations Containing Fractions

What is the procedure to remove the fractions in an equation before combining like terms?

Example 4:
Work through example 4 on page 2.2-6 and record the answer below.

Solve: $y+\frac{2}{3}=\frac{1}{5}$

Example 5:
Work through example 5 on page 2.2-7 and record the answer below. Check your answer by clicking on the link. If your answer is incorrect, watch the video to find your error.

Solve: $\frac{w+3}{2}-4=w+\frac{1}{3}$

Example 6:
Work through example 6 on page 2.2-9, using two techniques. For the first technique, clear the fractions before applying the properties of equality. For the second technique, do not clear the fractions before applying the properties of equality. Check your answer by clicking on the link. If your answer is incorrect, watch the video to find your error.

Solve: $\frac{5x}{2}-\frac{7}{8}=\frac{3}{4}x-\frac{11}{8}$

Technique 1

Technique 2

Sections 2.2 Objective 3: Solve Linear Equations Containing Decimals; Apply a General Strategy

What is the procedure to remove the decimals in an equation before combining like terms?

Example 7:
Work through example 7 on page 2.2-10 7 and record the answer below. Check your answer by clicking on the link.

Solve. $1.4x - 3.8 = 6$

Example 8:
Work through example 8 on page 2.2-11 and record the answer below. Check your answer by clicking on the link. If your answer is incorrect, watch the video to find your error.

Solve. $0.1x + 0.03(7 - x) = 0.05(7)$

Read and summarize the CAUTION statement on page 2.2-12.

What are the steps of **A General Strategy for Solving Linear Equations in One Variable**?

1.

2.

3.

4.

5.

6.

Section 2.2 Objective 4: Identify Contradictions and Identities

What are the three cases for the solution of a linear equation in one variable?

When no variable terms remain and a false statement results the equation is called a(n) ____________________, which means what?

When no variable terms remain and a true statement results the equation is called a(n) ____________________, which means what?

Example 9:
Study the solution for example 9 part a on page 2.2-16 and record the answer below. Complete part b on your own and check your answer by clicking on the link. If your answer is incorrect, watch the video to find your error.

Determine if the equation is a contradiction or an identity. State the solution set.

a. $3x + 2(x - 4) = 5x + 7$

b. $3(x - 4) = x + 2(x - 6)$

Section 2.2 Objective 5: Use Linear Equations to Solve Application Problems

Review the **Strategy for Solving Linear Equations in One Variable.**

Example 10:
Work through example 10 on page 2.2-17 and record the answer below.

The body surface area and weight of well proportioned infants and children are related by the equation

$$30S = W + 4$$

where S = the body surface area in square meters and W = weight in kilograms. Find the body surface area of a well-proportioned child that weighs 18 kg.

Example 11:
Work through example 11 on page 2.2-18 and record the answer below. Check your answer by clicking on the link. If your answer is incorrect, watch the video to find your error.

In the U.S., the average pounds of red meat eaten, M, is related to the average pounds of poultry eaten, P, by the equation

$$100M = 14000 - 42P$$

Determine the average amount of poultry eaten if the average amount of red meat eaten is 100.1 pounds.

Section 2.3 Guided Notebook

Section 2.3 Introduction to Problem Solving

Read the list of "THINGS TO KNOW" and review any concepts you are unfamiliar with.

Section 2.3 Objective 1: Translate Sentences into Equations
Complete the table as found on page 2.3-3.

Key Words That Translate to an Equal Sign			

Example 1:
Study the solutions for example 1 parts a and b on page 2.3-4 and record the answers below. Complete parts c and d on your own and check your answers by clicking on the link. If your answers are incorrect, watch the video to find your error.

Translate each sentence into an equation. Use x to represent each unknown number.

a. Fifty-two less than a number results in –21.

b. Three-fourths of a number, increased by 8, gives the number.

c. The difference of 15 and a number is the same as the sum of the number and 1.

d. If the sum of a number and 4 is multiplied by 2, the result will be 2 less than the product of 4 and the number.

Section 2.3 Objective 2: Use the Problem-Solving Strategy to Solve Direct Translation Problems

What is a **mathematical model**?

Section 2.3

What are the six steps of the **Problem-Solving Strategy for Applications of Linear Equations**?

1.

2.

3.

4.

5.

6.

Example 2:
Work through example 2 on page 2.3-7 and record the answer below.

Five times a number, increased by 17, is the same as 11 subtracted from the number. Find the number.

Example 3:
Work through example 3 on page 2.3-9 and record the answer below. Check your answer by clicking on the link. If your answer is incorrect, watch the video to find your error.

Four times the difference of twice a number and 5 results in the number increased by 50. Find the number.

<u>**Sections 2.3 Objective 3:** Solve Problems Involving Related Quantities</u>

For some problems, we need to find ______________________________ quantities that are related in some way.

Example 4:
Work through example 4 on page 2.3-10 and record the answer below.

The storage capacity of Deon's external hard drive is 32 times that of his jump drive, a small portable memory device. Together, his two devices have 264 gigabytes of memory. What is the memory size of each device?

Example 5:
Work through example 5 on page 2.3-11 and record the answer below. Check your answer by clicking on the link. If your answer is incorrect, watch the video to find your error.

Disney's *Toy Story* is 11 minutes shorter than its sequel *Toy Story 2*. *Toy Story 3* is 17 minutes longer than *Toy Story 2*. If the total running time for the three movies is 282 minutes, find the running time of each movie.

Section 2.3 Objective 4: Solve Problems Involving Consecutive Integers

What are **consecutive integers**?

Give examples of each of the following:

Three ***consecutive integers***.

Three ***consecutive even integers***.

Three ***consecutive odd integers***

What is the *general relationship* of each of the following?

Three consecutive integers
x, ________, ________

Three consecutive even integers
x, ________, ________

Three consecutive odd integers
x, ________, ________

Read and summarize the CAUTION statement on page 2.3-13.

Example 6:
Work through example 6 on page 2.3-14 and record the answer below.

The sum of two consecutive integers is 79. Find the two integers.

Example 7:
Work through example 7 on page 2.3-15 and record the answer below. Check your answer by clicking on the link. If your answer is incorrect, watch the video to find your error.

Three consecutive even integers add to 432. Find the three integers.

Section 2.3 Objective 5: Solve Problems Involving Value

What is the equation that relates value per item and the number of items?

Read and summarize the CAUTION statement on page 2.3-16.

Example 8:
Work through example 8 on page 2.3-16 and record the answer below. Check your answer by clicking on the link. If your answer is incorrect, watch the video to find your error.

Ethan's cell phone plan costs $34.99 per month for the first 700 minutes, plus $0.35 for each additional minute. If Ethan's bill is $57.39, how many minutes did he use?

Section 2.4 Guided Notebook

Section 2.4 Formulas

Read the list of "THINGS TO KNOW" and review any concepts you are unfamiliar with.

Section 2.4 Objective 1: Evaluate a Formula

What is the definition of a **formula**?

What is the **distance formula**?

Example 1:

Work through example 1 on page 2.4-3 and record the answer below.

A car travels at an average speed (rate) of 55 miles per hour for 3 hours. How far does the car travel?

Read and summarize the CAUTION statement on page 2.4-4.

What is the **Devine Formula** used for?

Example 2:

Study the solution for example 2 part a on page 2.4-4 and record the answer below. Complete part b on your own and check your answer by clicking on the link. If your answer is incorrect, watch the video to find your error.

Evaluate the Devine Formula to find the ideal body weight of each person described.

a. A man 72 inches tall

b. A woman 66 inches tall

Look at the two popup boxes on page 2.4-5 and write down the following formulas.

Perimeter

Area

Volume

Example 3:
Study the solution for example 3 part a on page 2.4-6 and record the answer below. Complete part b on your own and check your answer by clicking on the link. If your answer is incorrect, watch the video to find your error.

a. The top of a stainless steel sink is shaped like a square with each side measuring $15\frac{3}{4}$ inches long. How many inches of aluminum molding will be required to surround the outside of the sink?

b. A yield sign has the shape of a triangle with a base of 3 feet and a height of 2.6 feet. Find the area of the sign.

Section 2.4 Objective 2: Find the Value of a Non-isolated Variable in a Formula

Example 4:
Work through example 4 on page 2.4-8 and record the answer below.

The perimeter of a rectangle is given by the formula $P = 2l + 2w$, if $P = 84$ cm and $l = 26$ cm, find w.

What is the formula for **simple interest**? Be sure to identify the variables.

Example 5:
Study the solution for example 5 part a on page 2.4-9 and record the answer below. Complete part b on your own and check your answer by clicking on the link. If your answer is incorrect, watch the video to find your error.

Paige has invested \$15,000 in a certificate of deposit (CD) that pays 4% simple interest annually. If she earns \$750 in interest when the CD matures, how long has Paige invested the money?

What is the formula that relates Fahrenheit and Celsius measures of temperature?

Example 6:
Work through example 6 on page 2.4-10 and record the answer below. Check your answer by clicking on the link. If your answer is incorrect, watch the video to find your error.

During the month of February, the average high temperature in Montreal, QC is -4.6°C while the average high temperature in Phoenix, AZ is 70.7°F.

a. What is the equivalent Fahrenheit temperature in Montreal?

b. What is the equivalent Celsius temperature in Phoenix?

Sections 2.4 Objective 3: Solve a Formula for a Given Variable

What does it mean to **solve a formula for a given variable**?

Read and summarize the CAUTION statement on page 2.4-11.

Example 7:
Study the solutions for example 7 parts a and b on page 2.4-11 and record the answers below. Complete part c on your own and check your answer by clicking on the link. If your answer is incorrect, watch the video to find your error.

Solve each formula for the given variable.

a. Selling price: $S = C + M$ for M

b. Area of a triangle: $A = \frac{1}{2}bh$ for b

c. Perimeter of a rectangle: $P = 2l + 2w$ for l

Read and summarize the CAUTION statement on page 2.4-13

Section 2.4 Objective 4: Use Geometric Formulas to Solve Applications

Example 8:
Work through example 8 on page 2.4-14 and record the answer below.

An above-ground pool is shaped like a circular cylinder with diameter of 28 ft and a depth of 4.5 ft. If 1 $\text{ft}^3 \approx 7.5$ gal, how many gallons of water will the pool hold? Use $\pi = 3.14$ and round to the nearest thousand gallons.

Example 9:
Work through example 9 on page 2.4-15 and record the answers below. Check your answers by clicking on the link. If your answers are incorrect, watch the video to find your error.

Terrence wants to have a new floor installed in his living room, which measures 20 ft by 15 ft. Extending out 3 ft from one wall is a fireplace in the shape of a trapezoid with base lengths of 4 ft and 8 ft.

a. Find the area that needs flooring.

b. If the flooring costs $5.29 per square foot, how much will Terrence pay for the new floor? (Assume there is no wasted flooring.)

Example 10:
Work through example 10 on page 2.4-17 and record the answer below. Check your answers by clicking on the link. If your answers are incorrect, watch the video to find your error.

The roundabout intersection with the top view shown in the figure will be constructed using concrete pavement 9 inches thick. How many cubic yards of concrete will be needed for the roundabout? Use $\pi \approx 3.14$.

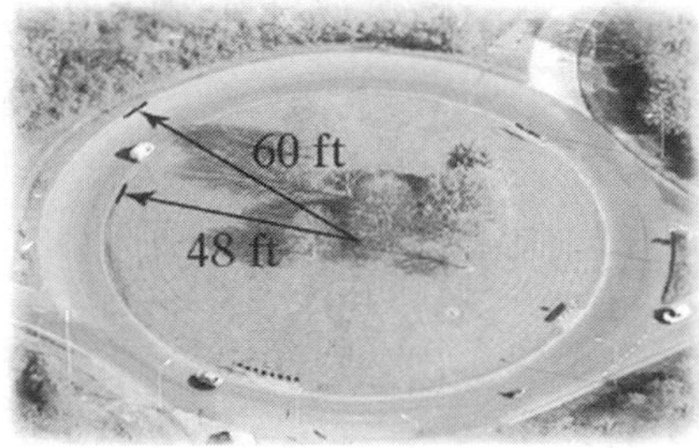

Section 2.5 Guided Notebook

Section 2.5 Geometry and Uniform Motion Problem Solving

Read the list of "THINGS TO KNOW" and review any concepts you are unfamiliar with.

Section 2.5 Objective 1: Solve Problems Involving Geometry Formulas

Example 1:
Work through example 1 on page 2.5-3 and record the answer below.

A green on a miniature golf course has a rectangular boundary. The length of the boundary is six feet longer than twelve times its width. If the perimeter is 103 feet, what are the dimensions of the green?

Example 2:
Work through example 2 on page 2.5-5 and record the answer below. Check your answer by clicking on the link. If your answer is incorrect, watch the video to find your error.

The blade of a canoe paddle is in the shape of an isosceles triangle so that two sides have the same length. The two common sides are each 4 inches longer than twice the length of the third side. If the perimeter is 48 inches, find the lengths of the sides of the blade.

Example 3:
Work through example 3 on page 2.5-6 and record the answer below. Check your answer by clicking on the link. If your answer is incorrect, watch the video to find your error.

The Triangle Drive In has been a local favorite for hamburgers in Fresno, CA since 1963. It is located on a triangular-shaped lot. One side of the lot is 4 meters longer than the shortest side, and the third side is 32 meters less than twice the length of the shortest side. If the perimeter of the lot is 180 meters, find the lengths of each side.

Example 4:
Work through example 4 on page 2.5-7 and record the answer below. Check your answer by clicking on the link. If your answer is incorrect, watch the video to find your error.

A bathtub is surrounded on three sides by a vinyl wall enclosure. The height of the enclosure is 20 inches less than three times the width, and the length is 5 inches less than twice the width. If the sum of the length, width, and height is 185 inches, what is the volume of the enclosed space?

Section 2.5 Objective 2: Solve Problems Involving Angles

What are **complementary angles**?

What are **supplementary angles**?

How do you find the following angles?

a. **Complement**

b. **Supplement**

Example 5:
Work through example 5 on page 2.5-9 and record the answer below.

Find the measure of each complementary angle in the following figure.

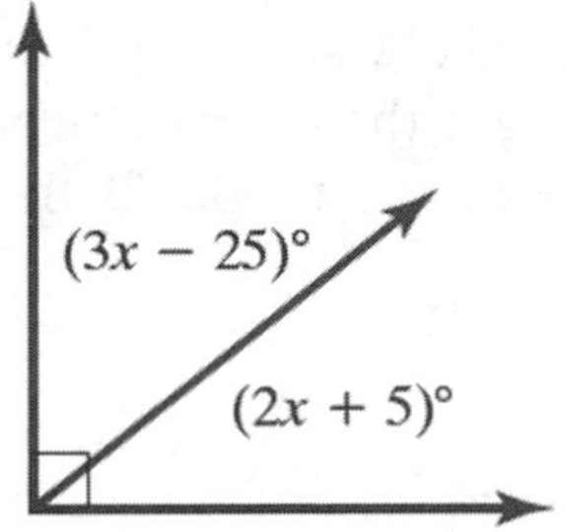

Example 6:
Work through example 6 on page 2.5-11 and record the answer below. Check your answer by clicking on the link. If your answer is incorrect, watch the video to find your error.

Find the measure of each supplementary angle in the following figure.

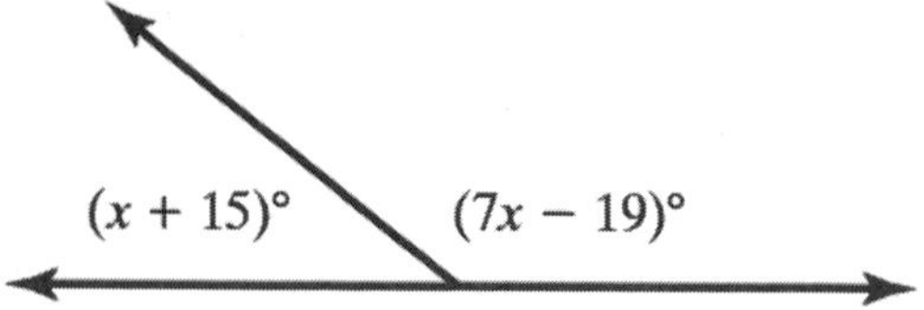

What is the **sum** of the angles of a triangle?

Example 7:
Work through example 7 on page 2.5-13 and record the answer below. Check your answer by clicking on the link. If your answer is incorrect, watch the video to find your error.

Triangle Park in Lexington, KY, has a roughly triangular shape such that the smallest angle measures 10 degrees less than the middle-sized angle. The largest angle measures 30 degrees less than twice the middle-sized angle. Find the measures of all three angles.

Section 2.5 Objective 3: Solve Problems Involving Uniform Motion

What are **uniform motion** problems?

Example 8:
Work through example 8 on page 2.5-14 and record the answer below.

In January 2010, the U.S government announced plans for the development of high-speed rail projects. A medium-fast passenger train leaves a station traveling 100 mph. Two hours later, a high-speed passenger train leaves the same station traveling 180 mph on a different track. How long will it take the high-speed train to be the same distance from the station as the medium-fast passenger train?

Section 2.5

Take notes on the **animation** on page 2.5-16.

Example 9:

Work through example 9 on page 2.5-16 and record the answer below. Check your answer by clicking on the link. If your answer is incorrect, watch the video to find your error.

Brennan provides in-home healthcare in a rural county and gets reimbursed for mileage. On one particular day he spent 4 hours driving to visit patients. His average speed is 50 mph on the highway but then slows to 30 mph when driving through towns. If he traveled five times as far on the highway as through towns, how far did he travel that day?

Read and summarize the CAUTION statement on page 2.5-18.

Section 2.6 Guided Notebook

Section 2.6 Percent and Mixture Problem Solving

Read the list of "THINGS TO KNOW" and review any concepts you are unfamiliar with.

Section 2.6 Objective 1: Solve Problems by Using a Percent Equation

What operation does the word "of" indicate?

What is the **general equation for percents**?

How is each of the following identified in percent problems?

Base

Percent

Amount

Read and summarize the CAUTION statement on page 2.6-4.

Example 1:
Study the solution for example 1 part a on page 2.6-4 and record the answer below. Complete parts b and c on your own and check your answers by clicking on the link. If your answers are incorrect, watch the video to find your error.

Use equations to solve each percent problem.

a. 32 is 40% of what number?

b. 145% of 78 is what number?

c. 8.2 is what percent of 12.5?

Example 2:
Work through example 2 on page 2.6-5 and record the answer below.

6% of a 128 fluid-ounce bottle of bleach is sodium hypochlorite. How many fluid ounces of sodium hypochlorite are in the bottle?

Section 2.6 Objective 2: Solve Percent Problems Involving Discount, Markups, and Sales Tax

What is a **discount**?

What are the formulas for **computing discounts**?

Example 3:
Work through example 3 on page 2.6-7 and record the answer below.

A furniture store is going out of business and cuts all prices by 55%. What is the sale price of a sofa with an original price of $1199?

What is a **markup**?

What are the formulas for **computing markups**?

Example 4:
Work through example 4 on page 2.6-8 and record the answer below. Check your answer by clicking on the link. If your answer is incorrect, watch the video to find your error.

A college book store sells all textbooks at a 30% markup over its cost. If the price marked on a biology textbook is $124.28, what was the cost of the book to the store? Round to the nearest cent.

What is **sales tax**?

What are the formulas for **computing sales tax**?

Example 5:
Work through example 5 on page 2.6-9 and record the answer below. See how to *check* your answer by clicking on the link.

Charlotte bought a pair of jeans priced at $51.99. When sales tax was added, she paid an overall price of $55.37. What was the tax rate? Round to the nearest tenth of a percent.

Sections 2.6 Objective 3: Solve Percent of Change Problems

What does **percent of change** describe?

What is a **percent of increase**? What is a **percent of decrease**?

What is the formula for **percent of increase**?

Example 6:
Work through example 6 on page 2.6-11 and record the answer below. See how to *check* your answer by clicking on the link.

Last year, 16,528 students attended City Community College. This year, enrollment increased by 3.2%. How many students attend City Community College this year? Round to the nearest whole student.

What is the formula for **percent of decrease**?

Example 7:
Work through example 7 on page 2.6-13 and record the answer below. Check your answer by clicking on the link. If your answer is incorrect, watch the video to find your error.

Prior to reorganization in 2010, General Motors (GM) had 91,000 U.S. employees. After the reorganization, GM had 68,500 U.S. employees. By what percent did the number of U.S. employees decrease?

Section 2.6

Read and summarize the CAUTION statement on page 2.6-13.

Section 2.6 Objective 4: Solve Mixture Problems

What does a 12% concentration mean?

What is the formula for a **Mixture Problem Equation**?

Take notes on the **animation** on page 2.6-14.

Example 8:
Work through example 8 on page 2.6-15 and record the answer below. Check your answer by clicking on the link. If your answer is incorrect, watch the video to find your error.

An organic cranberry-grape juice is 40% grape juice, while an organic fruit cocktail juice is 10% grape juice. If 8 ounces of the cranberry-grape juice are mixed with 22 ounces of the fruit cocktail juice, what is the mixed juice's concentration of grape juice?

Example 9:
Work through example 9 on page 2.6-17 and record the answer below. Check your answer by clicking on the link. If your answer is incorrect, watch the video to find your error.

How many milliliters of a 25% alcohol solution must be mixed with 10 mL of a 60% alcohol solution to result in a mixture that is 30% alcohol?

According to the CAUTION statement on page 2.6-18, why must the concentration of a mixture always be between the concentrations of the two mixed solutions?

Section 2.7 Guided Notebook

Section 2.7 Linear Inequalities in One Variable

Read the list of "THINGS TO KNOW" and review any concepts you are unfamiliar with.

Section 2.7 Objective 1: Write the Solution Set of an Inequality in Set-Builder Notation

What are **inequalities**?

Write down an example of **set-builder notation** and explain each part.

Example 1:
Work through example 1 on page 2.7-4 and record the answers below.

Write the solution set of each inequality in set-builder notation.

a. $x < 6$ b. $y \geq -3$ c. $2 < m \leq 9$

Section 2.7 Objective 2: Graph the Solution Set of an Inequality on a Number Line

In showing solutions on a number line, when is an open circle used and when is a closed circle used?

Example 2:
Study the solutions for example 2 parts a and b on page 2.7-5 and record the answer below. Complete parts c–f on your own and check your answers by clicking on the link. If your answers are incorrect, watch the video to find your error.

Graph each solution set on a number line.

a. $\{x|x \leq 0\}$ b. $\{x|-2 \leq x < 4\}$ c. $\{x|x > -1\}$

d. $\{x|3 < x < 7\}$ e. $\{x|-1 \leq x \leq 5\}$ f. $\{x|x \textit{ is any real number}\}$

Sections 2.7 Objective 3: Use Interval Notation to Express the Solution Set of an Inequality

What is the **lower bound** and **upper bound** of an interval?

When are parentheses used for the endpoints? When are square brackets used for the endpoints?

Complete the table as found on page 2.7-8.

Graph	Interval Notation	Set-Builder Notation
	(a,b)	
		$\{x \mid a \le x \le b\}$
	(a,b] [a,b)	
		$\{x \mid x > a\}$ $\{x \mid x < b\}$
		$\{x \mid x \ge a\}$ $\{x \mid x \le b\}$

Example 3:
Study the solutions for example 3 parts a and b on page 2.7-9. Complete parts c–f on your own and check your answers by clicking on the link. If your answers are incorrect, watch the video to find your error.

Write each solution set using interval notation.

c. x is less than 4

d. x is between -1 and 5, inclusive

e. $\{x \mid x \textit{ is any real number}\}$

f. $\{x \mid 8 > x \ge -3\}$

Section 2.7 Objective 4: Solve Linear Inequalities in One Variable

Example 4:
Study the solution for example 4 part a on page 2.7-11 and record the answer below. Complete part b on your own and check your answer by clicking on the link. If your answer is incorrect, watch the video to find your error.

Solve each inequality using the addition property of inequality. Write the solution set in interval notation and graph it on a number line.

a. $x+5>4$

b. $y-3\le 1$

What is the **Multiplication Property of Inequality?**

Example 5:
Study the solution for example 5 part a on page 2.7-15 and record the answer below. Complete part b on your own and check your answer by clicking on the link. If your answer is incorrect, watch the video to find your error.

Solve each inequality. Graph the solution set on a number line and write the solution set in interval notation.

a. $-5x < 15$

b. $\frac{m}{3} \geq 2$

Write the **Guidelines for Solving Linear Inequalities in One Variable**

1.

2.

3.

4.

5.

6.

Example 6:
Study the solution for example 6 part a on page 2.7-17 and record the answer below. Complete part b on your own and check your answer by clicking on the link. If your answer is incorrect, watch the video to find your error.

Solve each inequality. Graph the solution set on a number line and write the solution set in interval notation.

a. $-3x+2 \leq 8$

b. $6x-3 > 4x+9$

Example 8:
Work through example 8 on page 2.7-19 and record the answer below. Check your answer by clicking on the link. If your answer is incorrect, watch the video to find your error.

Solve the inequality $4 + 2(3 - x) > 3(2x + 7) + 5$. Write the solution set in set-builder notation, and graph it on a number line.

Example 9:
Work through example 9 on page 2.7-19 and record the answer below. Check your answer by clicking on the link. If your answer is incorrect, watch the video to find your error.

Solve the inequality $\frac{n}{3} - 4 > -\frac{n}{6} + 1$. Write the solution set in interval notation.

What are a **contradiction** and an **identity**?

Example 10:
Study the solution for example 10 part a on page 2.7-20 and record the answer below. Complete part b on your own. If your answer is incorrect, watch the video to find your error.

Solve the following inequalities. Write each solution set in interval notation.

a. $10 - 2(x + 1) > -5x + 3(x + 8)$ b. $2(5 - x) - 2 < 3(x + 3) - 5x$

Section 2.7 Objective 5: Solve Three-Part Inequalities
Study the technique to solve a three-part inequality.

Example 11:
Work through example 11 on page 2.7-21 and record the answer below.

Solve the inequality $4 \leq 3x - 2 < 7$. Write this solution set in interval notation and then graph it on a number line.

Example 12:
Work through example 12 on page 2.7-22 and record the answer below. Check your answer by clicking on the link. If your answer is incorrect, watch the video to find your error.

Solve the inequality $-1 < \frac{2x+1}{3} < 1$ and write its solution set in interval notation.

Section 2.7 Objective 6: Use Linear Inequalities to Solve Application Problems.

What are the steps in the **Strategy for Solving Application Problems Involving Linear Inequalities**?

1.

2.

3.

4.

5.

6.

Example 13:
Study the solution for example 13 on page 2.7-24.

Example 14:
Work through example 14 on page 2.7-25 and record the answer below. Watch the video to see the solution in full detail.

Lesley is planning a birthday party for her son at Incredible Pizza. The party will cost \$43 plus \$15 for each guest. If she does not want to spend more than \$300, what is the largest number of guests that can attend the party?

Section 2.8 Guided Notebook

Section 2.8 Compound Inequalities; Absolute Value Equations and Inequalities

Read the list of "THINGS TO KNOW" and review any concepts you are unfamiliar with.

Section 2.8 Objective 1: Find the Union and Intersection of Two Sets

Write down the definition of **Intersection**. What symbol is used?

Example 2:
Work through example 2 showing all steps below. Check your answer by clicking on the link. If your answer is incorrect watch the video to find your error.

Let $A = \{x | x > -2\}$ and $B = \{x | x \leq 5\}$. Find $A \cap B$, the intersection of two sets.

Write down the definition of **Union**. What symbol is used?

Example 4:
Work through example 4 showing all steps below. Check your answer by clicking on the link. If your answer is incorrect watch the video to find your error.

Let $A = \{x | x < -2\}$ and $B = \{x | x \geq 5\}$ Find $A \cup B$, the union of the two sets.

Section 2.8 Objective 2: Solve Compound Linear Inequalities in One Variable

Write down the **Guidelines for Solving Compound Linear Inequalities**.

1.

2.

3.

Example 6:
Study the solution for example 6, and record the answer below.

Solve $3x-5<-2$ and $4x+11\geq 3$. Graph the solution set and then write it in interval notation.

Example 7:
Work through example 7 showing all steps below. Click on the link to check your answer. If your answer is incorrect, watch the video to find your error.

Solve $9-4x<-7$ or $5x+6<3(x+2)$. Graph the solution set and then write it in interval notation.

Explain why the solution to $2x-3\le -1$ and $x-7\ge -3$ is a null set.

Explain why the solution to $10x+7>2$ or $3x-6\le 9$ is the set of real numbers.

Section 2.8 Objective 3: Solve Absolute Value Equations

Write down the **Absolute Value Equation Property**.

Example 9:
Work through example 9 showing all steps below. Check your answer by clicking on the link. If your answer is incorrect watch the video to find your error.

Solve. $|1-3x|=4$

Read and summarize the CAUTION statement on 2.8-17.

Write down the **Strategy for Solving Absolute Value Equations**.

1.

2.

3.

4.

Example 13:
Work through example 13 showing all steps below. Check your answer by clicking on the link. If your answer is incorrect watch the video to find your error.

Solve. $-3|2-m|+8=2$

<u>**Section 2.8 Objective 4:** Solve Absolute Value Inequalities</u>

Study Figures 6 and 7.

Write down the **Absolute Value Inequality Property**.

1.

2.

Example 14:
Work through example 14 showing all steps below. Check your answer by clicking on the link. If your answer is incorrect watch the video to find your error.

Solve. $|2m-1|\leq 5$

Example 15:
Work through example 15 showing all steps below. Check your answer by clicking on the link. If your answer is incorrect watch the video to find your error.

Solve. $|5x+1|>3$

Section 2.8

Read and summarize the CAUTION statement on 2.8-25.

Write down the **Strategy for Solving Absolute Value Inequalities**.

1.

2.

3.

Example 16:
Study the solution for example 16, and record the answer below.

Solve. $|4x-3|+2 \le 7$

Example 17:
Work through example 17 showing all steps below. Check your answer by clicking on the link. If your answer is incorrect watch the video to find your error.

Solve. $5|1-2x|-3>12$

Section 3.1 Guided Notebook

Section 3.1 The Rectangular Coordinate System

Read the list of "THINGS TO KNOW" and review any concepts you are unfamiliar with.

Section 3.1 Objective 1: Read Line Graphs

Why are graphs often used?

What are **line graphs**?

Click on the link on page 3.1-3 and write down the definition of **line segments**.

Example 1:
Study the solution for example 1 parts a and b on page 3.1-4 and record the answers below. Complete parts c–e on your own and check your answers by clicking on the link. If your answers are incorrect, watch the video to find your error.

The following line graph shows the average daily temperature in St. Louis, MO for each month.

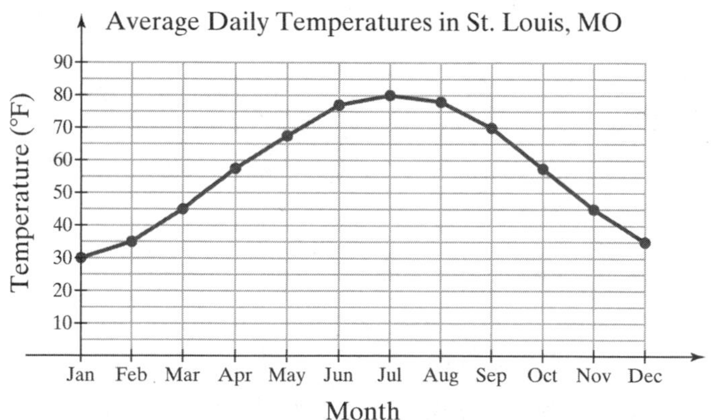

a. What is the average daily temperature in February?

b. What is the average daily temperature in November?

c. In what month is the average daily temperature 70° F?

d. Which month has the highest average daily temperature? What is the average daily temperature for that month?

e. In what months are the average daily temperatures above 65° F?

Section 3.1 Objective 2: Identify Points in the Rectangular Coordinate System

Watch the animation on page 3.1-6 and take notes below.

What is another name for the **rectangular coordinate system**?

What does the **Cartesian coordinate system** consist of?

Write down the definition of each of the following terms.

perpendicular

coordinate axes

***x*-axis**

***y*-axis**

origin

plane

coordinate plane

Cartesian plane

***xy*-plane**

quadrants

Draw a picture labeling each of the following: x-axis, y-axis, the four quadrants, origin

Write down the definition of each of the following terms.

point

ordered pair

x-coordinate

y-coordinate

abscissa

ordinate

Example 2:

Work through example 2 on page 3.1-8 and record the answers below. Check your answers by clicking on the link. If your answers are incorrect, watch the video to find your error.

Use an ordered pair to identify each point on the coordinate plane shown. State the quadrant or axis where each point lies.

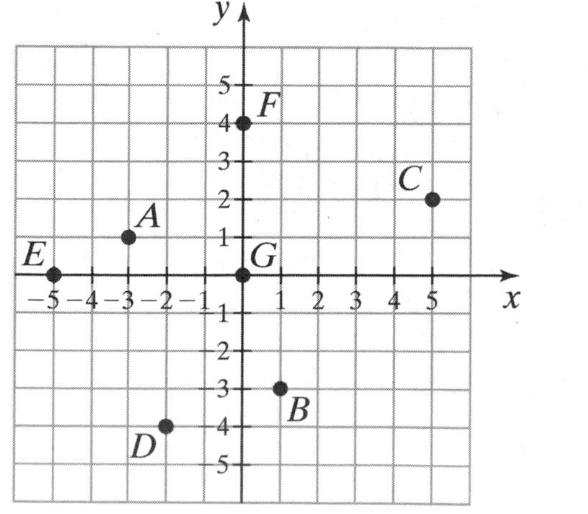

Point A

Point B

Point C

Point D

Point E

Point F

Point G

Section 3.1

Read and summarize the CAUTION statement on page 3.1-9. Draw a picture to demonstrate.

Section 3.1 Objective 3: Plot Ordered Pairs in the Rectangular Coordinate System.

To **plot** or ___________ an ordered pair, place a _____________ at its ______________ on the coordinate plane.

Example 3:
Work through example 3 on page 3.1-10 and record the answers below. Be sure to draw your graph large enough to view all the points. If your graph doesn't match the one shown on page 3.1-10, watch the video to find your error.

Plot each ordered pair on the coordinate plane. State the quadrant or axis where each point lies.

a. (2,4) b. (4,–5) c. (0,–2) d. (–3,-4) e. $\left(-\frac{7}{2},\frac{5}{2}\right)$ f. (1.5,0)

Section 3.1 Objective 4: Create Scatter Plots

What can ordered pairs be used to study? Give an example different from the book.

What are **paired data**?

What is a **scatter plot**?

Example 4:
Work through example 4 on 3.1-11 and record the answers below.

The table below shows the number of U.S. ethanol plants operating in the month of January for the years 2000–2010. List ordered pairs in the form (*year, number of plants*). Create a scatter plot of the paired data. Do the paired data show a trend? If so, what is the trend?

Year	Number of Plants
2000	54
2001	56
2002	61
2003	68
2004	72
2005	81
2006	95
2007	110
2008	139
2009	170
2010	189

Source: Renewable Fuels Association

Why is only Quadrant I needed for the answer?

According to the CAUTION statement on page 3.1-12, why do scatter plots typically involve only Quadrant I?

Section 3.2 Guided Notebook

Section 3.2 Graphing Linear Equations in Two Variables

Read the list of "THINGS TO KNOW" and review any concepts you are unfamiliar with.

Section 3.2 Objective 1: Determine If an Ordered Pair Is a Solution to an Equation

What is a **solution to an equation in one variable**?

What is a **Solution to an Equation in Two Variables**?

Example 1:
Study the solutions to example 1 parts a and b on page 3.2-4 and record the answers below. Complete parts c and d on your own and check your answers by clicking on the link. If your answers are incorrect, watch the video to find your error.

Determine if each ordered pair is a solution to the equation $x + 2y = 8$

a. $(-2,5)$ b. $(2,6)$ c. $\left(-11, \frac{3}{2}\right)$ d. $(0, 4)$

Section 3.2 Objective 2: Determine the Unknown Coordinate of an Ordered Pair Solution

What is an **ordered pair solution**?

Example 2:
Study the solution for example 2 part a on page 3.2-6 and record the answer below. Complete parts b and c on your own and check your answers by clicking on the link. If your answers are incorrect, watch the video to find your error.

Find the unknown coordinate so that each ordered pair satisfies $2x - 3y = 15$.

a. (6, ?) b. (? ,7) c. $\left(-\frac{5}{2}, ?\right)$

Section 3.2 Objective 3: Graph Linear Equation by Plotting Points

To find solutions to a linear equation select values for _____ and determine the corresponding value for _____ or select values for _____ and determine the corresponding value for _____.

What is the **graph of an equation in two variables**? Complete the two statements below.

a. It includes

b. It is a

To make such a graph:

1. ____________ several points that ________________ the ____________________

2. ____________ the points with a ________________ or ______________________

Write down a **Linear Equation in Two Variables (Standard Form)**.

How many points are required to determine a line?

Example 3:
Work through example 3 on page 3.2-10 and record your answer below.

Graph $3x - y = 2$ by plotting points.

Example 4:
Work through example 4 on page 3.2-14 and record the answer below. Check your answer by viewing figure 6b on page 3.2-15.

Graph $y = -\frac{1}{3}x + 2$ by plotting points.

Example 5:
Work through example 5 on page 3.2-16 and record the answers below. Check your answers by clicking on the link. If your answers are incorrect, watch the video to find your error.

Graph by plotting points.

a. $y = 2x$

b. $3x + 2y = 5$

Section 3.2 Objective 4: Find x- and y-Intercepts

What are **intercepts**?

What is a $\boldsymbol{y}$**-intercept**? What is the corresponding ordered pair?

What is an $\boldsymbol{x}$**-intercept**? What is the corresponding ordered pair?

Example 6:
Work through example 6 on page 3.2-18 and record the answer below.

Find the intercepts of the graph shown below. What are the x-intercepts? What are the y-intercepts?

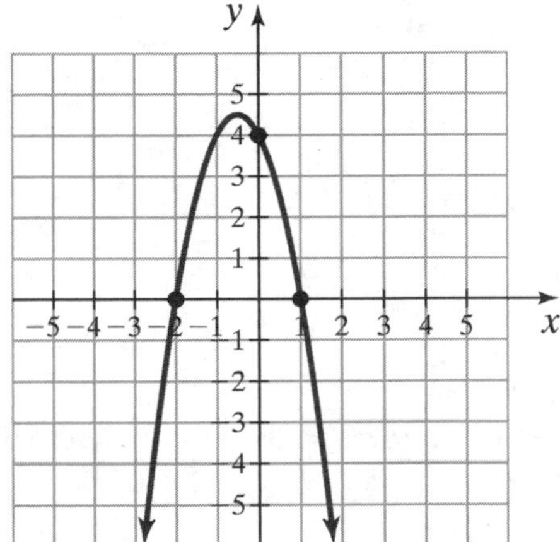

Determine how to find the $\boldsymbol{x}$**- and** $\boldsymbol{y}$**-Intercepts of a Graph Given an Equation**.

Example 7:
Study the solution for example 7 part a on page 3.2-20 and record the answer below. Complete part b on your own and check your answer by clicking on the link. If your answer is incorrect, watch the video to find your error.

Find the x- and y-intercepts for the graph of each equation.

a. $2x + y = 4$ b. $4x = 3y + 8$

Section 3.2 Objective 5: Graph Linear Equations Using Intercepts

Example 8:
Work through example 8 on page 3.2-21 and record the answer below.

Graph $3x - 2y = 6$ using intercepts.

To graph a linear equation with only one intercept, at least _______ more point(s) must be found.

Example 9:
Work through example 9 on page 3.2-23 and record the answer below. Check your answer by clicking on the link. If your answers are incorrect, watch the video to find your error.

Graph $2x = 5y$ using intercepts.

Section 3.2 Objective 6: Use Linear Equations to Model Data

Example 10:
Work through example 10 on page 3.2-24 and record the answers below. Check your answers by clicking on the link. If your answers are incorrect, watch the video to find your error.

The number of U.S. drive-in theaters can be modeled by the linear equation $y = -7.5x + 435$, where x is the number of years after 2000.

a. Sketch the graph of the equation for the year 2000 and beyond.

b. Find the missing coordinate for the ordered pair solution (?, 390).

c. Interpret the point from part (b).

d. Find and interpret the y-intercept.

e. What does the x-intercept represent in this problem?

Section 3.2 Objective 7: Graph Horizontal and Vertical Lines

Rewrite the equation $x = a$ as a linear equation in two variables in standard form.

Rewrite the equation $y = b$ as a linear equation in two variables in standard form.

What is the graph of the equation $x = a$?

What is the graph of the equation $y = b$?

Example 11:
Work through example 11 on page 3.2-25 and record the answer below.

Graph $y = 4$.

Example 12:
Work through example 12 on page 3.2-26 and record the answer below. Check your answer by clicking on the link. If your answers are incorrect, watch the video to find your error.

Graph $x = -3$.

Section 3.3 Guided Notebook

Section 3.3 Slope

Read the list of "THINGS TO KNOW" and review any concepts you are unfamiliar with.

<u>**Section 3.3 Objective 1:** Find the Slope of a Line Given Two Points</u>

What is a key feature of a line?

How is the **slant** or **steepness** of a line measured?

Write down the definition of **slope**.

What does a slope of $\frac{2}{3}$ mean?

What is the **slope formula**? Be sure to include the diagram.

Example 1:
Work through example 1 on page 3.3-7 and record the answer below.

Find the slope of the line containing the points (–2,-4) and (0,1). Include a graph of the line.

Read and summarize the CAUTION statement on page 3.3-8.

Example 2:
Work through example 2 on page 3.3-8 and record the answer below.

Find the slope of the line containing the points (–2,4) and (1,–3). Include a graph of the line.

Summarize the information on **Positive versus Negative Slope** from page 3.3-10.

Example 3:
Study the solution for example 3 part a on page 3.3-10 and record the answer below. Complete part b on your own and check your answer by clicking on the link. If your answer is incorrect, watch the video to find your error.

Find the slope of the line containing the given points. Simplify if possible.

a. (–6,–1) and (4,5)

b. (1,5) and (3,–1)

<u>**Section 3.3 Objective 2:** Find the Slopes of Horizontal and Vertical Lines</u>

Example 4:
Work through example 4 on page 3.3-11 and record the answers below. Watch the video for a detailed solution.

Find the slope of the line containing the given points. Simplify if possible. Include a graph of each line.

a. (–3,2) and (1,2)

b. (4,2) and (4,–5)

Summarize the information on **Slope of Horizontal and Vertical Lines** from page 3.3-13.

According to the CAUTION statement on page 3.3-13, why should the term "no slope" be avoided?

Draw and summarize the relationship between the slope and the graph of a linear equation as shown on page 3.3-13.

Section 3.3 Objective 3: Graph a Line Using the Slope and a Point

To sketch the graph of a line using the slope and a given point, we____________________

____

____.

Example 5:
Work through example 5 on page 3.3-14 and record the answer below.

Graph the line that has slope $m = \frac{3}{2}$ and passes through the point (1,–2).

Example 6:
Work through example 6 on page 3.3-15 and record the answer below. Check your answer by clicking on the link. If your answer is incorrect, watch the video to find your error.

Graph the line that has slope m = –3 and passes through the point (2,–1).

Section 3.3 Objective 4: Find and Use the Slopes of Parallel and Perpendicular Lines

What are **parallel lines**? Include a picture.

Write down the information on **Parallel Lines** found on page 3.3-17.

What are **perpendicular lines**? Include a picture.

What does **opposite reciprocals** mean?

Write down the information on **Perpendicular Lines** found on page 3.3-19.

Example 7:
Work through example 7 on page 3.3-20 and record the answers below.

Line l_1 has a slope of $m_1 = -\frac{4}{5}$

a. If line l_2 is parallel to l_1, what is its slope?

b. If line l_2 is perpendicular to l_1, what is its slope?

Example 8:
Work through example 8 on page 3.3-21 and record the answers below. Check your answers by clicking on the link. If your answers are incorrect, watch the video to find your error.

a. Graph a line l_2 that is parallel to l_1 and passes through the point (3,–2)

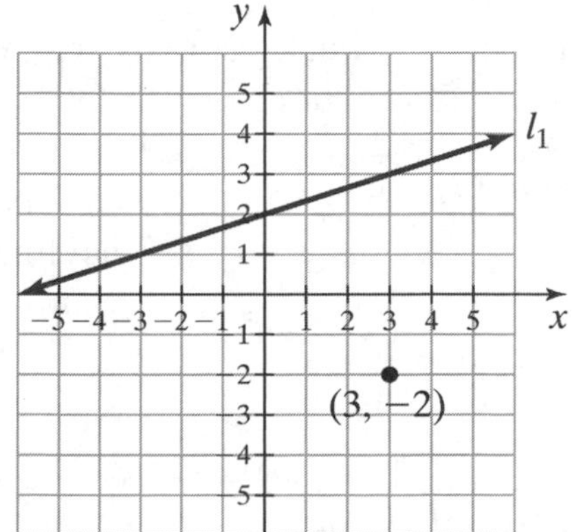

b. Graph a line l_3 that is perpendicular to l_1 and passes through the point (3,–2)

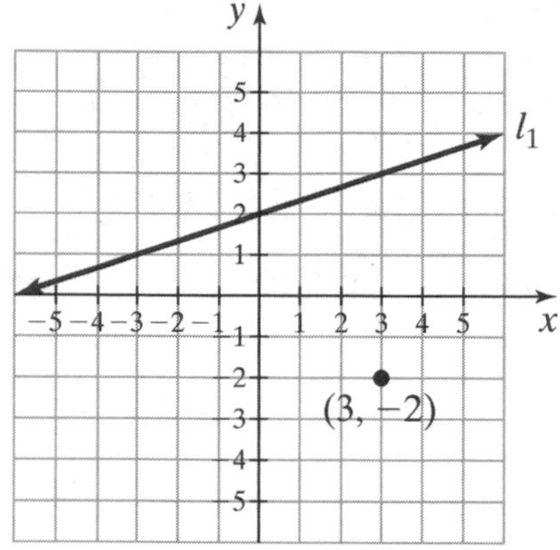

Section 3.3 Objective 5: Use Slope in Applications

What is **grade**?

What is the slope of a roof called?

Example 9:
Work through example 9 on page 3.3-22 and record the answer below.

A standard wheelchair ramp should rise no more than 1 foot vertically for every 12 feet horizontally. Find the grade of this ramp. Round to the nearest tenth of a percent.

What is **average rate of change**? Why is it called this in many applications?

Example 10:
Work through example 10 on page 3.3-24 and record the answer below. Watch the video for a detailed solution.

The average tuition and fees for U.S. public two-year colleges were \$2130 in 1999. The average tuition and fees were \$2540 in 2009. Find and interpret the slope of the line connecting the points (1999, 2130) and (2009, 2540).

Section 3.4 Guided Notebook

Section 3.4 Equations of Lines

Read the list of "THINGS TO KNOW" and review any concepts you are unfamiliar with.

Section 3.4 Objective 1: Determine the Slope and y-Intercept from a Linear Equation

To determine the slope and y-intercept directly from an equation what must be done first?

What is **Slope-Intercept Form**?

Read and summarize the CAUTION statement on page 3.4-4.

Example 1:
Study the solution to example 1 part a on page 3.4-5 recording the answer below. Complete part b on your own and check your answer by clicking on the link.

Find the slope and y-intercept of the given line.

a. $2x + y = 3$

b. $4x - 3y = 6$

Example 2:
Work through example 2 on page 3.4-6 and record the answers below. Check your answers by clicking the link. If your answers are incorrect, watch the video to find your error.

Find the slope and y-intercept of the given line.

a. $4x - 10y = 0$

b. $y = 4$

Section 3.4 Objective 2: Use the Slope-Intercept Form to Graph a Linear Equation

Study the animation on page 3.4-7.

Example 3:
Work through example 3 on page 3.4-7 and record the answer below.

Graph the equation $y = \frac{3}{5}x - 2$ using the slope and y-intercept.

Example 4:
Work through example 4 on page 3.4-9 and record the answer below. Check your answer by clicking on the link. If your answer is incorrect, watch the video to find your error.

Graph the equation $2x + 3y = 9$ using the slope and y-intercept.

Section 3.4 Objective 3: Write the Equation of a Line Given Its Slope and *y*-Intercept

If the slope and *y*-intercept is given, how do you write the equation of the line?

Example 5:
Work through example 5 on page 3.4-10 and record the answers below.

Write an equation of the line with the given slope and *y*-intercept.

a. slope -4; *y*-intercept 3

b. slope $\frac{2}{5}$; *y*-intercept -7

Section 3.4 Objective 4: Write the Equation of a Line Given Its Slope and a Point on the Line

Example 6:
Work through example 6 on page 3.4-12 and record the answer below.

Write the equation of the line that has slope 2 and passes through the point (4,7).

What is the **Point-Slope Form**?

Example 7:
Work though example 7 on page 3.4-13 and record the answer below. Check your answer by clicking the link. If your answer is incorrect, watch the video to find your error.

Use the point-slope form to determine the equation of the line that has slope $-\frac{3}{4}$ and passes through the point (2,–5). Write the equation in slope-intercept form.

Section 3.4

Example 8:
Work through example 8 on page 3.4-15 and record the answers below.

Write the equation of a line that passes through the point (–3,2) and has the given slope.

a. $m = 0$

b. undefined slope

Section 3.4 Objective 5: Write the Equation of a Line Given Two Points

Given two points on a line, we can find the equation of the line by

1.

2.

Example 9:
Work though example 9 on page 3.4-16 and record the answer below. Check your answer by clicking the link. If your answer is incorrect, watch the video to find your error.

Write the equation of the line passing through the points (–4,1) and (2,4). Write your answer in slope-intercept form.

Complete the following table as found on page 3.4-17

	Slope Average rate of change
	Point-Slope Form Slope is ____ and _______ is a point on the line
$y = mx + b$	
$Ax + By = C$	
	Horizontal Line Slope is _________, and y-intercept is ______
$x = a$	

Section 3.4 Objective 6: Determine the Relationship Between Two Lines

Parallel lines have the ____________ slope but _______________ y-intercepts.

Perpendicular lines have ____slopes.

Coinciding Lines have the ___________ slope and the ___________y-intercept.

Two lines with different slopes will __________________.

What are **only intersecting** lines?

Complete the following table as found on page 3.4-18.

	Parallel Lines	**Coinciding Lines**	**Intersecting and Perpendicular Lines**	**Only Intersecting Lines (not Perpendicular)**
Slopes are				
***y*-Intercepts are**				

Example 10:
Study the solution for example 10 part a on page 3.4-19 and record the answer below. Complete parts b–d on your own and check your answers by clicking on the link. If your answers are incorrect, watch the video to find your error.

For each pair of lines, determine if the lines are parallel, perpendicular, coinciding, or only intersecting.

a. $3y = -2y + 7$
   $3x - 2y = 8$

b. $y = -3x + 1$
   $6x + 2y = 2$

c. $4x - 5y = 15$
   $y = \frac{4}{5}x + 1$

d. $3x - 4y = 2$
   $x + 2y = -12$

Section 3.4 Objective 7: Write the Equation of a Line Parallel or Perpendicular to a Given Line

Example 11:
Study the solution for example 11 part a on page 3.4-20 and record the answer below. Complete part b on your own and check your answer by clicking the link. If your answer is incorrect, watch the video to find your error.

Write the equation of the line that passes through the point $(6,-5)$ and is

a. perpendicular to $6x - 2y = -1$

b. parallel to $y = -2x + 4$

Complete the following table as found on page 3.4-22.

If you are given……	**Then do this……**
Slope and *y*-intercept	
Slope and a point	
	Find the slope. Then use this slope and either point in the slope-intercept form to find b.
Point and slope =0 (horizontal line)	
	Use the *x*-coordinate of the point to write $x = \text{a}$
	Find the slope of the given line. Use this slope and the point in the slope-intercept form to find *b*.
Point and perpendicular line in equation form.	

Read the CAUTION statement on page 3.4-22.

<u>**Section 3.4 Objective 8:** Use Linear Equations to Solve Applications</u>

Example 12:
Work through example 12 on page 3.4-23 and record the answers below. Watch the video for the detailed solutions.

If attendance at professional football games is 17 million in a given year then the corresponding attendance at college football games is 31 million. Increasing attendance at professional football games to 25 million increases attendance at college football games to 55 million.

a. Assume that the relationship between professional football attendance (in millions) and college football attendance (in millions) is linear. Find the equation of the line that describes this relationship. Write your answer in slope-intercept form.

b. Use your equation from part (a) to estimate the attendance at college football games if the attendance at professional football games is 21 million

Section 3.5 Guided Notebook

Section 3.5 Linear Inequalities in Two Variables

Read the list of "THINGS TO KNOW" and review any concepts you are unfamiliar with.

Section 3.5 Objective 1: Determine If an Ordered Pair Is a Solution to a Linear Inequality in Two Variables

What is the solution set for a linear inequality in one variable?

Where is the solution set of a linear inequality in one variable typically graphed?

What is a **Linear Inequality in Two Variables?**

When is an ordered pair a **solution to a linear inequality in two variables**?

Example 1:
Study the solution to example 1 part a on page 3.5-4 and record the answer below. Complete parts b and c on your own and check your answers by clicking on the link. If your answers are incorrect, watch the video to find your error.

Determine if the given ordered pair is a solution to the inequality $2x - 3y < 6$.

a. $(-1,-2)$ b. $(4,-1)$ c. $(6,2)$

Section 3.5 Objective 2: Graph a Linear Inequality in Two Variables

To find solutions to $x + y < 2$ or $x + y > 2$ what equation do we start with? Graph that equation below. To get more details, click on the link on page 3.5-5.

Section 3.5

What is a **half-plane**?

Does the upper half-plane represent the solutions to $x + y < 2$ or $x + y > 2$. Why?

Does the lower half-plane represent the solutions to $x + y < 2$ or $x + y > 2$. Why?

What is a **boundary line**?

What are the **Steps for Graphing Linear Inequalities in Two Variables**?

1.

2.

3.

Read and summarize the **Note** on page 3.5-7

Example 2:
Study the solution for example 2 part a on page 3.5-7 and record the answer below. Complete parts b and c on your own and check your answers by clicking on the link. If your answers are incorrect, watch the video to find your error.

Graph each inequality.

a. $3x - 4y \leq 8$

b. $y > 3x$

c. $y < -2$

According to the CAUTION statement on 3.5-10, why can't a test point belong to the boundary line?

Section 3.5 Objective 3: Solve Applications Involving Linear Inequalities in Two Variables

Example 3:
Work through example 3 on 3.5-11 and record the answers below. Watch the video for a detailed solution.

A piggy bank contains only nickels and dimes with a total value of less than \$9. Let n = the number of nickels and d = the number of dimes.

a. Write an inequality describing the possible numbers of coins in the bank.

b. Graph the inequality. Because n and d must be whole numbers, restrict the graph to Quadrant I.

c. Could the piggy bank contain 90 nickels and 60 dimes?

Read and summarize the CAUTION statement on page 3.5-12.

Section 4.1 Guided Notebook

Section 4.1 Solving Systems of Linear Equations by Graphing

Read the list of "THINGS TO KNOW" and review any concepts you are unfamiliar with.

Section 4.1 Objective 1: Determine If an Ordered Pair Is a Solution to a System of Linear Equations in Two Variables

What is the definition of a **System of Linear Equations in Two Variables**? Give three examples.

What is the definition of the **Solution to a System of Linear Equations in Two Variables**?

Read and summarize the CAUTION statement on page 4.1-5.

Example 1:
Work through example 1 on page 4.1-5 and record the answers below. Check your answers by clicking on the link. If your answers are incorrect, watch the video to find your error.

Determine if each ordered pair is a solution to the following system:

$$\begin{cases} 2x+3y=12 \\ x+2y=7 \end{cases}$$

a. (–3, 6)

b. (3,2)

Section 4.1 Objective 2: Determine the Number of Solutions to a System Without Graphing

What are the three possible outcomes when two linear equations are graphed? Include a sketch of each.

Write down the definitions for the following terms.

Consistent

Inconsistent

Dependent

Independent

Describe the slopes and *y*-intercepts of two lines with the following number of solutions.

One solution

No solutions

Infinite number of solutions

Example 2:
Study the solution for example 2 part a on page 4.1-8 and record the answer below. Complete parts b and c on your own and check your answers by clicking on the link. If your answers are incorrect, watch the video to find your answer.

Determine the number of solutions to each system without graphing.

a. $\begin{cases} y = 3x - 4 \\ 6x + 3y = 8 \end{cases}$　　b. $\begin{cases} 2x - 4y = \dfrac{8}{3} \\ 3x - 6y = 4 \end{cases}$　　c. $\begin{cases} 5x - 2y = 3 \\ -\dfrac{5}{2}x + y = 7 \end{cases}$

<u>Sections 4.1 Objective 3:</u> <u>Solve Systems of Linear Equations by Graphing</u>

What are the three methods for solving systems of linear equations in two variables?

There is no need to solve which systems by graphing? Why?

What are the steps for **Solving Systems of Linear Equations in Two Variables by Graphing**?

1.

2.

3.

Read and summarize the CAUTION statement on page 4.1-11.

Example 3:
Work through example 3 on page 4.1-12 and record the answer below.

Solve the following system by graphing:

$$\begin{cases} y = 2x+1 \\ y = -x+4 \end{cases}$$

Example 4:
Work through example 4 on page 4.1-14 and record the answer below. Watch the video to see the fully worked solution.

Solve the following system by graphing:

$$\begin{cases} 3x + y = -2 \\ x + y = 2 \end{cases}$$

Section 4.2 Guided Notebook

Section 4.2 Solving Systems of Linear Equations by Substitution

Read the list of "THINGS TO KNOW" and review any concepts you are unfamiliar with.

Section 4.2 Objective 1: Solve Systems of Linear Equations by Substitution

Summarize the **substitution method** as found on page 4.2-4.

What are the steps for **Solving Systems of Linear Equations in Two Variables by Substitution**?

1.

2.

3.

4.

Read and summarize the CAUTION statement on page 4.2-4.

Example 1:
Work through example 1 on page 4.2-5 and record the answer below.

Use the substitution method to solve the following system:

$$\begin{cases} 4x+2y=10 \\ y=3x-10 \end{cases}$$

Example 2:
Work through example 2 on page 4.2-7 and record the answer below. Check your answer by clicking on the link. If your answer is incorrect, watch the video to find your error.

Use the substitution method to solve the following system:

$$\begin{cases} 3x + \ \ y = -6 \\ 2x - \dfrac{1}{3}y = 2 \end{cases}$$

Example 3:
Work through example 3 on page 4.2-8 and record the answer below. Check your answer by clicking on the link. If your answer is incorrect, watch the video to find your error.

Solve the following system:

$$\begin{cases} 4x + 3y = 7 \\ x + 9y = -1 \end{cases}$$

Section 4.2

Example 4:
Work through example 4 on page 4.2-9 and record the answer below. Check your answer by clicking on the link. If your answer is incorrect, watch the video to find your error.

Use the substitution method to solve the following system:

$$\begin{cases} 6x-3y=-33 \\ 2x+4y=4 \end{cases}$$

Section 4.2 Objective 2: Solve Special Systems by Substitution

What are the three possible outcomes when two linear equations are graphed? Include a sketch of each.

Write down the definitions for the following terms.

Consistent

Inconsistent

Dependent

Independent

Describe the slopes, *y*-intercepts, and number of solutions of two lines described below. Also state if they are consistent, inconsistent, dependent, and/or independent. Show an example for each.

Intersecting Lines

Parallel Lines

Coinciding Lines

When solving algebraically, how are the following situations recognized?

1. The system is independent and consistent.
2. The system is independent and inconsistent.
3. The system is dependent and consistent.

Example 5:
Work through example 5 on page 4.2-11 and record the answer below

Use the substitution method to solve the following system:

$$\begin{cases} 2x+10y=8 \\ x+5y=4 \end{cases}$$

Read and summarize the CAUTION statement on page 4.2-15.

Example 6:
Work through example 6 on page 4.2-15 and record the answer below.

Use the substitution method to solve the following system:

$$\begin{cases} 3x - y = -1 \\ -12x + 4y = 8 \end{cases}$$

Example 7:
Work through example 7 on page 4.2-17 and record the answers below. Check your answers by clicking on the link. If your answer is incorrect, watch the video to find your error.

Use the substitution method to solve the following system:

a. $\begin{cases} \frac{1}{4}x + y = 5 \\ x + 4y = 8 \end{cases}$

b. $\begin{cases} -2.4x + 1.5y = -3 \\ 0.8x - 0.5y = 1 \end{cases}$

Section 4.3 Guided Notebook

Section 4.3 Solving Systems of Linear Equations by Elimination

Read the list of "THINGS TO KNOW" and review any concepts you are unfamiliar with.

Section 4.3 Objective 1: Solve Systems of Linear Equations by Elimination

Summarize the **elimination method** as found on page 4.3-3. Include the **Logic for the Elimination Method**.

What is another name for the elimination method?

Example 1:
Work through example 1 on page 4.3-3 and record the answer below.

Solve the following system.

$$\begin{cases} x+y=8 \\ x-y=-2 \end{cases}$$

To eliminate a variable, the coefficients of the variable in the two equations must be

____________________. How can we make this happen?

Example 2:
Work through example 2 on page 4.3-6 record the answer below. Check your answer by clicking on the link.

Solve the following system.

$$\begin{cases} x - y = -4 \\ x + 2y = 5 \end{cases}$$

What are the five steps for **Solving Systems of Linear Equations in Two Variables by Elimination**?

1.

2.

3.

4.

5.

Read and summarize the CAUTION statement on page 4.3-8.

Example 3:
Work through example 3 on page 4.3-9 and record the answer below. If your answer is incorrect, watch the video to find your error.

Solve the following system:

$$\begin{cases} x - 3y = -9 \\ 5x + 4y = -7 \end{cases}$$

Example 4:
Work through example 4 on page 4.3-11 and record the answer below. Check your answer by clicking on the link. If your answer is incorrect, watch the video to find your error.

Use the elimination method to solve the following system.

$$\begin{cases} -3x + 4y = 7 \\ 5x + 6y = 1 \end{cases}$$

Example 5:
Work through example 5 on page 4.3-12 and record the answer below. Check your answer by clicking on the link. If your answer is incorrect, watch the video to find your error.

Use the elimination method to solve the following system.

$$\begin{cases} 5x - 6y = 20 \\ 4x + 9y = 16 \end{cases}$$

Example 6:
Work through example 6 on page 4.3-13 and record the answer below. Check your answer by clicking on the link. If your answer is incorrect, watch the video to find your error.

Use the elimination method to solve the following system.

$$\begin{cases} x - \frac{3}{5}y = \frac{4}{5} \\ \frac{1}{2}x + 3y = -\frac{9}{5} \end{cases}$$

Section 4.3 Objective 2: Solve Special Systems by Elimination

When solving a system by elimination, an ______________________________ system will lead to a ______________________________ and a ______________________________ system will lead to an ______________________________.

Example 7:
Study the solution for example 7 part a on page 4.3-14 and record the answer below. Complete part b on your own and check your answer by clicking on the link. If your answer is incorrect, watch the video to find your error.

Use the elimination method to solve each system.

a. $\begin{cases} 3x + y = 6 \\ 6x + 2y = 4 \end{cases}$

b. $\begin{cases} 2x - 8y = 6 \\ 3x - 12y = 9 \end{cases}$

Section 4.4 Guided Notebook

Section 4.4 Applications of Linear Systems

Read the list of "THINGS TO KNOW" and review any concepts you are unfamiliar with.

Section 4.4 Objective 1: Solve Related Quantity Applications Using Systems

What are the six steps for the **Problem-Solving Strategy for Applications Using Systems of Linear Equations**?

1.

2.

3.

4.

5.

6.

Example 1:
Work through example 1 on page 4.4-4 and record the answer below.

The storage capacity of Deon's external hard drive is 32 times that of his jump drive, a small portable memory device. Together, his two devices have 264 gigabytes of memory. What is the memory size of each device?

Write down the solution process for the same problem from section 2.3, example 4. Use the link found on page 4.4-5.

Example 2:
Work through example 2 on page 4.4-6 and record the answer below. Check your answers by clicking on the link. If your answers are incorrect, watch the video to find your error.

The sum of the ages of Ben and his younger sister Annie is 18 years. The difference of their ages is 4 years. What is the age of each child?

Section 4.4 Objective 2: Solve Geometry Applications Using Systems

Example 3:
Work through example 3 on page 4.4-7 and record the answer below. Check your answers by clicking on the link. If your answers are incorrect, watch the video to find your error.

The display panel of a graphing calculator has the shape of a rectangle with a perimeter of 264 millimeters. If the length of the display panel is 18 millimeters longer than the width, find its dimensions.

The measures of two **complementary angles** add to __________, while the measures of two

supplementary angles add to ___________.

Example 4:
Work through example 4 on page 4.4-8 and record the answer below. Check your answers by clicking on the link. If your answers are incorrect, watch the video to find your error.

Find the measures of two supplementary angles if the measure of the larger angle is 20 degrees less than three times the measure of the smaller angle.

Section 4.4 Objective 3: Solve Uniform Motion Applications Using Systems

Example 5:
Work through example 5 on page 4.4-9 and record the answer below. Check your answers by clicking on the link. If your answers are incorrect, watch the video to find your error.

Shawn is training for the Dirty Duo running-and-bicycling race. During a three hour training session, his total distance cycling and running was 33 miles. If he cycled at a rate of 18 miles per hour and ran at a rate of 6 miles per hour, how much time did he spend doing each activity?

When motions work together the rates are ______________, but when they work against

each other, the rates are _______________.

Example 6:
Work through example 6 on page 4.4-11 and record the answers below. Check your answers by clicking on the link. If your answers are incorrect, watch the video to find your error.

A jet plane travels 1950 miles in 3.9 hours going with the wind. On the return trip, the plane must fly into the wind and the travel time increases to 5 hours. Find the speed of the jet plane in still air and the speed of the wind. Assume the wind speed is the same for both trips.

Section 4.4 Objective 4: Solve Mixture Applications Using Systems

Example 7:
Work through example 7 on page 4.4-14 and record the answers below. Check your answers by clicking on the link. If your answers are incorrect, watch the video to find your error.

A shipping company delivered 160 packages one day. The cost of regular delivery is $6.50, and the cost for express delivery is $17.50. Total shipping revenue for the day was $1513. How many of each kind of delivery were made?

Example 8:
Work through example 8 on page 4.4-16 and record the answers below. Check your answers by clicking on the link. If your answers are incorrect, watch the video to find your error.

A chemist needs eight liters of a 50% alcohol solution but only has a 30% solution and an 80% solution available. How many liters of each solution should be mixed to form the needed solution?

Example 9:
Work through example 9 on page 4.4-19 and record the answers below. Check your answers by clicking on the link. If your answers are incorrect, watch the video to find your error.

Logan and Payton went to Culver's for lunch. Logan ate two Butterburgers with cheese and a small order of fries for a total of 1801 calories. Payton ate one Butterburger with cheese and two small orders of fries for a total of 1313 calories. How many calories are in a Culver's Butterburger with cheese? How many calories are in a small order of fries?

Section 4.5 Guided Notebook

Section 4.5 Systems of Linear Inequalities

Read the list of "THINGS TO KNOW" and review any concepts you are unfamiliar with.

Section 4.5 Objective 1: Determine If an Ordered Pair Is a Solution to a System of Linear Inequalities in Two Variables

What is the definition of a **System of Linear Inequalities in Two Variables**? Show two examples.

What is the definition of a **Solution to a System of Linear Inequalities in Two Variables**?

Example 1:
Study the solution for example 1 part a on page 4.5-4 and record the answer below. Complete parts b and c on your own and check your answers by clicking on the link. If your answers are incorrect, watch the video to find your answer.

Determine if each ordered pair is a solution to the following system of inequalities.

$$\begin{cases} 2x + y \geq -3 \\ x - 4y \leq 12 \end{cases}$$

a. (4,2) b. (2,–5) c. (0,–3)

Section 4.5 Objective 2: Graph Systems of Linear Inequalities

What is the **graph of a system of linear inequalities in two variables**?

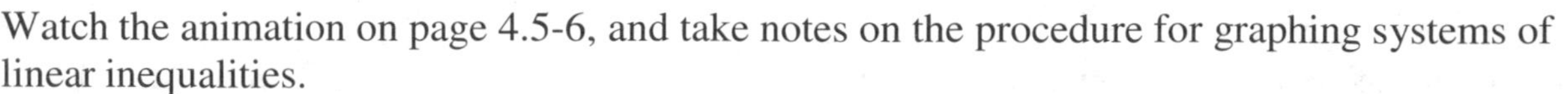

Watch the animation on page 4.5-6, and take notes on the procedure for graphing systems of linear inequalities.

What are the **Steps for Graphing Systems of Linear Inequalities**?

1.

2.

Read and summarize CAUTION statement on page 4.5-6.

Example 2:
Work through example 2 on page 4.5-7 and record the answer below. Watch the video for a detailed solution.

Graph the system of linear inequalities.

$$\begin{cases} 2x + y \geq -3 \\ x - 4y \leq 12 \end{cases}$$

Read and summarize the CAUTION statement on page 4.5-8.

Example 3:
Work through example 3 on page 4.5-9 and record the answer below. Check your answer by clicking on the link. If your answer is incorrect, watch the video to find your error.

Graph the system of linear inequalities.

$$\begin{cases} x+y<4 \\ x-2y<-2 \end{cases}$$

Read and summarize the CAUTION statement on page 4.5-9.

How many solutions does a system of linear inequalities typically have?

What is an **inconsistent system of inequalities**?

Example 4:
Work through example 4 on page 4.5-10 and record the answer below. Watch the video for a detailed solution.

Graph the system of linear inequalities.

$$\begin{cases} y \le -\dfrac{1}{3}x-3 \\ y > -\dfrac{1}{3}x+2 \end{cases}$$

Example 5:
Work through example 5 on page 4.5-12 and record the answer below. Watch the video for a detailed solution.

Graph the system of linear inequalities.

$$\begin{cases} x-3y>6 \\ x\geq 1 \end{cases}$$

Read and summarize the CAUTION statement on page 4.5-13.

Explain the solution region for the system below by viewing the link on page 4.5-14.

$$\begin{cases} x\geq 0 \\ y\geq 0 \end{cases}$$

Example 6:
Work through example 6 on page 4.5-14 and record the answer below. Watch the video for a detailed solution.

Graph the system of linear inequalities.

$$\begin{cases} x+y<4 \\ y>0 \\ 2x-y>-4 \end{cases}$$

Read and summarize the CAUTION statement on page 4.5-15.

Section 4.5 Objective 3: Solve Applications Involving Systems of Linear Inequalities

Example 7:
Work through example 7 on page 4.5-16 and record your answers below. Watch the video for a detailed solution.

Savannah is planning a barbeque for her family and friends. She will spend \$150 or less to buy hamburger patties that cost \$3 per pound and boneless chicken breasts that cost \$5 per pound. To limit waste, she will purchase at most 40 pounds of meat all together. Also, the amount of hamburger and chicken purchased must be non-negative. A system of linear inequalities that models this situation is

$$\begin{cases} 3h+5c \le 150 \\ h+c \le 40 \\ h \ge 0 \\ c \ge 0 \end{cases}$$

Where h = pounds of hamburger patties and c = pounds of chicken breasts.

a. Graph the system of linear inequalities.

b. Can Savannah purchase 20 pounds of hamburger patties and 15 pounds of chicken breasts for the barbeque?

c. Can Savannah purchase 10 pounds of hamburger patties and 30 pounds of chicken breasts for the barbeque?

Section 4.6 Guided Notebook

Section 4.6 Systems of Linear Equations in Three Variables

Read the list of "THINGS TO KNOW" and review any concepts you are unfamiliar with.

Section 4.6 Objective 1: Determine If an Ordered Triple Is a Solution to a System of Linear Equations in Three Variables

Write down the definition of a **System of Linear Equations in Three Variables**. Provide one example.

Example 1:

Study the solution for example 1 part a, and record the answer below. Complete part b on your own and check your answer by clicking on the link. If your answer is incorrect watch the video to find your error.

Determine if each ordered triple is a solution to the given system: $\begin{cases} 3x + y - 2z = 4 \\ 2x - 2y + 3z = 9 \\ x + y - z = 5 \end{cases}$

a. $(3,9,7)$

b. $(2,-4,-1)$

Section 4.6 Objective 2: Solve Systems of Linear Equations in Three Variables

Study the six possible solution scenarios for a system of linear equations in three variables.

Write down the **Guidelines for Solving a System of Linear Equations in Three Variables by Elimination**.

1.

2.

3.

4.

5.

6.

Example 3:

Work through example 3 showing all steps below. Check your answer by clicking on the link. If your answer is incorrect, watch the video to find your error.

Solve the following system: $\begin{cases} 2x + y = 13 \\ 3x - 2y + z = 8 \\ x + 2y - 3z = 5 \end{cases}$

Example 4:
Work through example 4 showing all steps below. Check your answer by clicking on the link. If your answer is incorrect, watch the video to find your error.

Solve the following system: $\begin{cases} \frac{1}{2}x + y + \frac{2}{3}z = 2 \\ \frac{3}{4}x + \frac{5}{2}y - 2z = -7 \\ x + 4y + 2z = 4 \end{cases}$

Example 5:
Study the solution for example 5, which is an example of an **inconsistent system**.

Solve the following system: $\begin{cases} x - y + 2z = 5 \\ 3x - 3y + 6z = 15 \\ -2x + 2y - 4z = 7 \end{cases}$

Example 6:
Study the solution for example 6, which is an example of a **dependent system**.

Solve the following system: $\begin{cases} x - y + 2z = 5 \\ 3x - 3y + 6z = 15 \\ -2x + 2y - 4z = -10 \end{cases}$

Section 4.6

Read and summarize the CAUTION statement on 4.6-23.

Section 4.6 Objective 3: Use Systems of Linear Equations in Three Variables to Solve Application Problems

Review the six-step **Problem-Solving Strategy for Applications Using Systems of Linear Equations.**

Example 9:
Work through example 9 showing all steps below. Check your answer by clicking on the link. If your answer is incorrect, watch the video to find your error.

Wendy ordered 30 T-shirts online for her three children. Small T-shirts cost $4 each, medium T-shirts cost $5 each, and large T-shirts are $6 each. She spent $40 more for the large T-shirts than for the small T-shirts. Wendy's total bill was $154. How many T-shirts of each size did she buy?

Section 5.1 Guided Notebook

Section 5.1 Exponents

Read the list of "THINGS TO KNOW" and review any concepts you are unfamiliar with.

Section 5.1 Objective 1: Simplify Exponential Expressions Using the Product Rule

Write down the definitions for the following terms.

Exponential expression

Base

Exponent

What is the **Product Rule for Exponents**?

Example 1:
Study the solutions for example 1 parts a and b on page 5.1-5 and record the answers below. Complete parts c and d on your own and check your answers by clicking on the link. If your answers are incorrect, watch the video to find your error.

Use the product rule to simplify each expression.

a. $5^4 \cdot 5^6$ b. $x^5 \cdot x^7$ c. $y^3 \cdot y$ d. $b^3 \cdot b^5 \cdot b^4$

Read and summarize the CAUTION statement on page 5.1-5.

Example 2:
Study the solutions for example 2 parts a and b on page 5.1-6 and record the answers below. Complete part c on your own and check your answer by clicking on the link. If your answer is incorrect, watch the video to find your error.

Simplify using the product rule.

a. $(4x^2)(7x^3)$ b. $(m^4n^2)(m^3n^6)$ c. $(-3a^5b^3)(-8a^2b)$

<u>**Section 5.1 Objective 2:** Simplify Exponential Expressions Using the Quotient Rule</u>

What is the **Quotient Rule for Exponents**?

Example 3:
Study the solutions for example 3 parts a and b on page 5.1-8 and record the answers below. Complete parts c and d on your own and check your answers by clicking on the link. If your answers are incorrect, watch the video to find your error.

Use the quotient rule to simplify each expression

a. $\dfrac{t^9}{t^5}$ b. $\dfrac{7^5}{7^3}$ c. $\dfrac{y^{24}}{y^{15}}$ d. $\dfrac{(-4)^{14}}{(-4)^{11}}$

Example 4:
Study the solutions for example 4 parts a and b on page 5.1-9 and record the answers below. Complete part c on your own and check your answer by clicking on the link. If your answer is incorrect, watch the video to find your error.

Simplify using the quotient rule.

a. $\dfrac{15x^6}{3x^2}$ b. $\dfrac{a^4b^9c^5}{a^2b^3c}$ c. $\dfrac{4m^6n^7}{12m^5n^2}$

Section 5.1 Objective 3: Use the Zero-Power Rule

What is the **Zero-Power Rule**?

Example 5:
Work through example 5 on page 5.1-11 and record the answers below.

Simplify using the zero-power rule.

a. 6^0 b. $(-3)^0$ c. -3^0 d. $(2x)^0$ e. $2x^0$

Section 5.1 Objective 4: Use the Power-to-Power Rule

What is the **Power-to-Power Rule**?

Example 6:
Study the solution for example 6 part a on page 5.1-14 and record the answer below. Complete part b on your own and check your answer by clicking on the link. If your answer is incorrect, watch the video to find your error.

Simplify using the power-to-power rule.
a. $(y^5)^6$ b. $[(-2)^3]^5$

Section 5.1 Objective 5: Use the Product-to-Power Rule

What is the **Product-to-Power Rule**?

Example 7:
Study the solutions for example 7 parts a and b on page 5.1-16 and record the answers below. Complete parts c and d on your own and check your answers by clicking on the link. If your answers are incorrect, watch the video to find your error.

Simplify using the product-to-product rule.

a. $(mn)^8$ b. $(x^2y)^5$ c. $(3y)^4$ d. $(-4p^5q^3)^2$

Section 5.1 Objective 6: Use the Quotient-to-Power Rule

What is the **Quotient-to-Power Rule**?

Example 8:
Study the solutions for example 8 parts a and b on page 5.1-18 and record the answers below. Complete parts c and d on your own and check your answers by clicking on the link. If your answers are incorrect, watch the video to find your error.

Simplify using the quotient-to-power rule.

a, $\left(\frac{m}{n}\right)^9$ b. $\left(\frac{x^2}{y^5}\right)^4$ c. $\left(\frac{x}{2}\right)^5$ d. $\left(\frac{3x^2}{5y^4}\right)^3$

Section 5.1 Objective 7: Simplify Exponential Expressions Using a Combination of Rules

What are the four conditions for an exponential expression to be considered **simplified**?

-
-
-
-

Write down the **Rules for Exponents**.

1.

2.

3.

4.

5.

6.

Example 9:
Study the solutions for example 9 parts a and b on page 5.1-20 and record the answers below. Complete parts c and d on your own and check your answers by clicking on the link. If your answers are incorrect, watch the video to find your error.

Simplify using the rules for exponents.

a. $(c^3)^5(c^2)^6$ b. $\left(\dfrac{15x^8y^5}{3x^6y}\right)^2$ c. $(-2w^3z^2)(-2wz^2)^4$ d. $\dfrac{(4m^2n^0)(2n^3)^2}{8mn^5}$

Section 5.2 Guided Notebook

Section 5.2 Introduction to Polynomials

Read the list of "THINGS TO KNOW" and review any concepts you are unfamiliar with.

Section 5.2 Objective 1: Classify Polynomials

What is a **simplified term?**

What is the definition of a **Monomial**?

Which of the terms from the interactive video on page 5.2-3 is a monomial? Why?

Write down the definitions for the following terms.

Polynomial

Terms of the polynomial

Simplified polynomial

Polynomials in one variable

How many terms do each of the following have? Give an example of each.

Monomial

Binomial

Trinomial

Polynomial

Example 1:
Work through example 1 on page 5.2-5 and record the answers below. Click on the link to check your answers.

Classify each polynomial as a monomial, binomial, trinomial, or none of these.

a. $5x - 7$ b. $\frac{1}{3}x^2$ c. $5x^3 - 7x^2 + 4x + 1$ d. $-2x^3 - 5x^2 + 8x$

Section 5.2 Objective 2: Determine the Degree and Coefficient of a Monomial

What is the definition of the **Degree of a Monomial**?

What is the degree of any constant term?

What is the definition of the **Coefficient of a Monomial**?

Example 2:
Study the solutions for example 2 parts a–c on page 5.2-6 and record the answers below. Complete parts d–f on your own and check your answers by clicking on the link. If your answers are incorrect, watch the video to find your error.

Determine the coefficient and degree of each monomial.

a. $4.6x^3$ b. $7x$ c. x^2y^4 d. 12 e. $\frac{3}{4}x^2yz^3$ f. $-2xyz^7$

Section 5.2 Objective 3: Determine the Degree and Leading Coefficient of a Polynomial

What is the definition of the **Degree of a Polynomial**?

What does it mean for a polynomial to be written in **descending order**?

What is **standard form** for polynomials?

What is the definition of the **Leading Coefficient of a Polynomial in One Variable**?

Example 3:
Study the solution for example 3 part a on page 5.2-9 and record the answer below. Complete part b on your own and check your answer by clicking on the link. If your answer is incorrect, watch the video to find your error.

Write each polynomial in standard form. Then find its degree and leading coefficient.

a. $4.2m - 3m^2 + 1.8 - 7m^3$

b. $\frac{2}{3}x^3 - 3x^2 + 5 - x^4 + \frac{1}{4}x$

Section 5.2 Objective 4: Evaluate a Polynomial for a Given Value

Example 4:
Study the solutions for example 4 parts a and b on page 5.2-10 and record the answers below. Complete parts c and d on your own and check your answers by clicking on the link. If your answers are incorrect, watch the video to find your error.

Evaluate the polynomial $x^3 + 3x^2 + 4x - 5$ for the given values of x.

a. $x = -2$

b. $x = 0$

c. $x = 2$

d. $x = \frac{5}{2}$

Section 5.2 Objective 5: Simplify Polynomials by Combining Like Terms

Example 5:
Study the solutions for example 5 parts a and b on page 5.2-12 and record the answers below. Complete parts c–e on your own and check your answers by clicking on the link. If your answers are incorrect, watch the video to find your error.

Simplify each polynomial by combining like terms.

a. $3x^2+8x-4x+2$

b. 2.3x – 3 – 5x + 8.4

c. $2x+3x^2-6+x^2-2x+9$

d. $\frac{2}{3}x^2+\frac{1}{5}x-\frac{1}{10}x-\frac{1}{6}x^2+\frac{1}{4}$

e. $6x^3+x^2-7$

Section 5.3 Guided Notebook

Section 5.3 Adding and Subtracting Polynomials

Read the list of "THINGS TO KNOW" and review any concepts you are unfamiliar with.

Section 5.3 Objective 1: Add Polynomials

What is the procedure for **Adding Polynomials**?

Example 1:
Work through example 1 on page 5.3-3 and record the answer below.

Add: $(2x + 8) + (7x - 3)$.

Example 2:
Study the solution for example 2 part a on page 5.3-5 and record the answer below. Complete parts b and c on your own and check your answers by clicking on the link. If your answers are incorrect, watch the video to find your error.

Add.

a. $(y^2 + 3y + 7) + (y^2 - 3y - 2)$

b. $(10p^3 + 7p - 13) + (5p^2 - 4p)$

c. $(3m^3 + m^2 - 8) + (2m^3 - 4m^2 + 3m) + (5m^2 + 4)$

Section 5.3 Objective 2: Find the Opposite of a Polynomial

How to you find the **opposite of a polynomial?**

Example 3:
Work through example 3 on page 5.3-6 and record the answers below.

Find the opposite of each polynomial

a. $x^2 + 6x + 8$ b. $8y - 27$ c. $-m^3 - 5m^2 + m + 7$

Section 5.3 Objective 3: Subtract Polynomials

What is the procedure for **Subtracting Polynomials**?

Example 4:
Study the solution for example 4 part a on page 5.3-8 and record the answer below. Complete part b on your own and check your answer by clicking on the link. If your answer is incorrect, watch the video to find your error.

Subtract.

a. $(9x + 13) - (6x - 4)$ b. $(3a^2 + 5a - 8) - (-2a^2 + a - 7)$

Section 5.4 Guided Notebook

Section 5.4 Multiplying Polynomials

Read the list of "THINGS TO KNOW" and review any concepts you are unfamiliar with.

Section 5.4 Objective 1: Multiply Monomials

What is the procedure for **Multiplying Monomials**?

Example 1:
Study the solution for example 1 part a on page 5.4-3 and record your answer below. Complete parts b and c on your own and check your answers by clicking on the popup.

Multiply.

a. $(6x^5)(7x^2)$ b. $\left(-\frac{3}{4}x^2\right)\left(-\frac{2}{9}x^8\right)$ c. $(3x^2)(-0.2x^3)$

Section 5.4 Objective 2: Multiply a Polynomial by a Monomial

What is the procedure for **Multiplying Polynomials by Monomials**?

Example 2:
Study the solution for example 2 part a on page 5.4-4 and record your answer below. Complete part b on your own and check your answer by clicking on the popup.

Multiply.

a. $3x(4x-5)$ b. $-4x^2(3x^2+x-7)$

Example 3:
Work through example 3 on page 5.4-5 and record your answers below. Check your answers by clicking on the link. If your answers are incorrect, watch the video to find your error.

Multiply.

a. $\frac{1}{2}x^2(4x^2-6x+2)$ b. $0.25x^3(6x^3-10x^2+4x-7)$

Section 5.4 Objective 3: Multiply Two Binomials

What is the procedure for **Multiplying Two Binomials**?

Example 4:
Study the solutions for example 4 parts a and b on page 5.4-6 and record your answers below. Complete part c on your own and check your answer by clicking on the link. If your answer is incorrect, watch the video to find your error.

Multiply using the distributive property twice.

a. $(x+3)(x+2)$ b. $(x+6)(x-2)$ c. $(x-4)(x-5)$

Read and summarize the CAUTION statement on page 5.4-8.

Take notes on the animation on page 5.4-8 illustrating how to use the **FOIL method**.

Example 5:
Study the solution for example 5 part a on page 5.4-8 and record your answer below. Complete parts b and c on your own and check your answers by clicking on the link. If your answers are incorrect, watch the video to find your error.

Multiply using the FOIL method.

a. $(x-4)(2x+3)$ b. $\left(\frac{1}{2}x-6\right)(3x-4)$ c. $(5x+7)(4x+3)$

Read and summarize the CAUTION statement on page 5.4-10.

Section 5.4 Objective 4: Multiply Two or More Polynomials

Take notes on the animation on page 5.4-11 and explain how to **Multiply Two or More Polynomials**.

Example 6:
Study the solution for example 6 part a on page 5.4-11 and record your answer below. Complete part b on your own and check your answer by clicking on the link. If your answer is incorrect, watch the video to find your error.

Multiply.

a. $(x+2)(2x^2-7x+3)$ b. $(y^2+2y-9)(2y^2-4y+7)$

Example 7:
Work through example 7 on page 5.4-12 and record your answers below. Check your answers by clicking on the link. If your answers are incorrect, watch the video to find your error.

Multiply.

a. $-4x(2x-1)(x+3)$

b. $(x-1)(x+3)(3x-2)$

Section 5.5 Guided Notebook

Section 5.5 Special Products

Read the list of "THINGS TO KNOW" and review any concepts you are unfamiliar with.

Section 5.5 Objective 1: Square a Binomial Sum

What is a **binomial sum**?

What is the **Square of a Binomial Sum Rule**?

Read and summarize the CAUTION statement on page 5.5-3.

Example 1:
Study the solution for example 1 parts a and b on page 5.5-4 and record the answers below. Complete parts c and d on your own and check your answers by clicking on the link. If your answers are incorrect, watch the video to find your error.

Multiply.

a. $(x+7)^2$ b. $(0.2m+1)^2$ c. $\left(z^2+\frac{1}{4}\right)^2$ d. $\left(10y+\frac{2}{5}\right)^2$

Section 5.5 Objective 2: Square a Binomial Difference

What is a **binomial difference**?

What is the **Square of a Binomial Difference Rule**?

Read and summarize the CAUTION statement on page 5.5-5.

Example 2:
Study the solutions for example 2 parts a and b on page 5.5-6 and record the answers below. Complete parts c and d on your own and check your answers by clicking on the link. If your answers are incorrect, watch the video to find your error.

Multiply.

a. $(x-3)^2$ b. $\left(2z-\frac{1}{6}\right)^2$ c. $(w^3-0.7)^2$ d. $(5p-1.2)^2$

What are **perfect square trinomials**?

<u>**Section 5.5 Objective 3:** Multiply the Sum and Difference of Two Terms</u>

What are **conjugates**?

What is the **Sum and Difference of Two Terms Rule (Product of Conjugates Rule)**?

Example 3:
Study the solution for example 3 parts a and b on page 5.5-9 and record the answers below. Complete parts c and d on your own and check your answers by clicking on the link. If your answers are incorrect, watch the video to find your error.

Multiply.

a. $(x+4)(x-4)$ b. $\left(5y+\frac{1}{2}\right)\left(5y-\frac{1}{2}\right)$ c. $(8-x)(8+x)$ d. $(3z^2+0.5)(3z^2-0.5)$

Write down the three **Special Product Rules for Binomials**.

Section 5.6 Guided Notebook

Section 5.6 Negative Exponents and Scientific Notation

Read the list of "THINGS TO KNOW" and review any concepts you are unfamiliar with.

Section 5.6 Objective 1: Use the Negative Power Rule

What is the definition of a **Negative Exponent**?

Example 1:
Study the solutions for example 1 parts a–c on page 5.6-4 and record the answers below. Complete parts d–f on your own and check your answers by clicking on the link. If your answers are incorrect, watch the video to find your error.

Write each expression with positive exponents. Then simplify if possible.

a. x^{-4} b. 2^{-3} c. $7x^{-3}$ d. $(-2)^{-4}$ e. -3^{-2} f. $2^{-1}+3^{-1}$

Read and summarize the CAUTION statement on page 5.6-5.

What is the **Negative-Power Rule**?

Example 2:
Study the solutions for example 2 parts a–c on page 5.6-6 and record the answers below. Complete parts d–f on your own and check your answers by clicking on the link. If your answers are incorrect, watch the video to find your error.

Write each expression with positive exponents. Then simplify if possible.

a. $\dfrac{1}{y^{-5}}$ b. $\dfrac{1}{6^{-2}}$ c. $\dfrac{3}{4t^{-7}}$ d. $\dfrac{-8}{q^{-11}}$ e. $\dfrac{m^{-9}}{n^{-4}}$ f. $\dfrac{5^{-3}}{2^{-4}}$

Section 5.6 Objective 2: Simplify Expressions Containing Negative Exponents Using a Combination of Rules

What are the four requirements for **simplified exponential expressions**?

-
-
-
-

What are the seven **Rules for Exponents**?

1.
2.
3.
4.
5.
6.
7.

Example 3:

Study the solutions for example 3 parts a and b on page 5.6-9 and record the answers below. Complete parts c and d on your own and check your answers by clicking on the link. If your answers are incorrect, watch the video to find your error.

Simplify.

a. $(9x^{-5})(7x^{2})$ b. $(p^{-4})^{2}$ c. $\frac{52m^{-4}}{13m^{-10}}$ d. $(w^{-1}z^{3})^{-4}$

Example 4:
Study the solutions for example 4 parts a and b on page 5.6-10 and record the answers below. Complete parts c and d on your own and check your answers by clicking on the link. If your answers are incorrect, watch the video to find your error.

Simplify.

a. $\dfrac{(3xz)^{-2}}{(2yz)^{-3}}$ b. $\left(\dfrac{10}{x}\right)^{-3}$ c. $\dfrac{\left(2a^5b^{-6}\right)^3}{4a^{-1}b^5}$ d. $\left(\dfrac{-5xy^{-3}}{x^{-2}y^5}\right)^4$

Section 5.6 Objective 3: Convert a Number from Standard From to Scientific Notation

What is **Scientific Notation**?

What is the procedure for **Converting from Standard Form to Scientific Notation**?

1.

2.

Example 5:
Study the solutions for example 5 parts a and b on page 5.6-14 and record the answers below. Complete parts c and d on your own and check your answers by clicking on the link. If your answers are incorrect, watch the video to find your error.

Write each number in scientific notation.

a. 56,800,000,000,000,000 b. 0.0000000467

c. 0.00009012 d. 200,000,000

Section 5.6 Objective 4: Convert a Number from Scientific Notation to Standard Form

What is the procedure for **Converting from Standard Form to Scientific Notation**?

1.

2.

Example 6:
Study the solutions for example 6 parts a and b on page 5.6-16 and record the answers below. Complete parts c and d on your own and check your answers by clicking on the link. If your answers are incorrect, watch the video to find your error.

Write each number in standard form.

a. 4.98×10^{-5} b. 9.4×10^{7} c. -3.015×10^{9} d. 1.203×10^{-4}

Section 5.6 Objective 5: Multiply and Divide with Scientific Notation

Example 7:
Study the solutions for example 7 parts a and b on page 5.6-17 and record the answers below. Complete parts c and d on your own and check your answers by clicking on the link. If your answers are incorrect, watch the video to find your error.

Perform the indicated operations. Write your results in scientific notation..

a. $\left(1.8\times10^{5}\right)\left(3\times10^{8}\right)$ b. $\dfrac{2.16\times10^{12}}{4.5\times10^{3}}$

c. $\left(-7.4\times10^{9}\right)\left(6.5\times10^{-4}\right)$ d. $\dfrac{5.7\times10^{-3}}{7.5\times10^{-7}}$

Section 5.7 Guided Notebook

Section 5.7 Dividing Polynomials

Read the list of "THINGS TO KNOW" and review any concepts you are unfamiliar with.

Section 5.7 Objective 1: Divide Monomials

What is the procedure for **Dividing Monomials**?

Example 1:
Work through example 1 on page 5.7-3 and record the answers below.

Divide

a. $\frac{32x^7}{4x^3}$ b. $\frac{9y^4}{45y^4}$ c. $\frac{60y}{5y^4}$

Section 5.7 Objective 2: Divide a Polynomial by a Monomial

What is the procedure for **Dividing Polynomials by Monomials**?

Example 2:
Study the solution for example 2 part a on page 5.7-5 and record the answer below. Complete part b on your own and check your answer by clicking on the link. If your answer is incorrect, watch the video to find your error.

Divide.

a. $\frac{12x^3-28x^2}{4x^2}$ b. $\left(9m^5-15m^4+18m^3\right)\div 3m^3$

Example 3:
Work through example 3 on page 5.7-7 and record the answer below.

Divide $\dfrac{54t^3 - 12t^2 - 24t}{6t^2}$

Section 5.7 Objective 3: Divide Polynomials Using Long Division

Read and summarize the CAUTION statement on page 5.7-8.

What is the **Process for Polynomial Long Division**?

1.

2.

3.

4.

Example 4:
Work through example 4 on page 5.7-9 and record the answer below.

Divide $\left(2x^2 + x - 15\right) \div \left(x + 3\right)$

Example 5:
Work through example 5 on page 5.7-12 and record the answer below. Check your answer by clicking on the link. If your answer is incorrect, watch the video to find your error.

Divide $\dfrac{x^2+26x-6x^3-12}{2x-3}$

Example 6:
Work through example 6 on page 5.7-14 and record the answer below. Check your answer by clicking on the link. If your answer is incorrect, watch the video to find your error.

Divide $\dfrac{3t^3-11t-12}{t+4}$

Section 5.8 Guided Notebook

Section 5.8 Polynomials in Several Variables

Read the list of "THINGS TO KNOW" and review any concepts you are unfamiliar with.

Section 5.8 Objective 1: Determine the Degree of a Polynomial in Several Variables

What is a **Polynomial in Several Variables**?

Example 1:
Study the solution for example 1 part a on page 5.8-3 and record your answer below. Complete part b on your own and check your answer by viewing the popup,

Determine the coefficient and degree of each term; then find the degree of the polynomial.

a. $2x^3y - 7x^2y^3 + xy^2$

b. $3x^2yz^3 - 4xy^3z + xy^2z^4$

Section 5.8 Objective 2: Evaluate Polynomials in Several Variables

Example 2:
Study the solution for example 2 part a on page 5.8-5 and record your answer below. Complete part b on your own and check your answer by clicking on the link. If your answer is incorrect, watch the video to find your error.

a. Evaluate $3x^2y - 2xy^3 + 5$ for $x = -2$ and $y = 3$.

b. Evaluate $-a^3bc^2 + 5a^2b^2c - 2ab$ for $a = 2$, $b = -1$ and $c = 4$.

Section 5.8 Objective 3: Add or Subtract Polynomials in Several Variables

We add polynomials in several variables by____

____.

Example 3:
Study the solutions for example 3 parts a and b on page 5.8-6 and record your answers below. Complete parts c and d on your own and check your answers by clicking on the link. If your answers are incorrect, watch the video to find your error.

Add or subtract as indicated.

a. $\left(2x^2+3xy-7y^2\right)+\left(4x^2-xy+11y^2\right)$

b. $\left(4a^2-3ab+2b^2\right)-\left(6a^2-5ab+7b^2\right)$

c. $\left(7x^4+3x^3y^3-2xy^3+5\right)+\left(2x^4-x^3y^3+8xy^3-10\right)$

d. $\left(10x^3y+2x^2y^2-5xy^3-8\right)-\left(6x^3y+x^2y^2-3xy^3\right)$

Section 5.8 Objective 3: Multiply Polynomials in Several Variables

We multiply polynomials in several variables ______________________________________

____.

Example 4:
Work through example 4 on page 5.8-8 and record the answer below.

Multiply: $5xy^2\left(4x^2-3xy+2y^2\right)$

Section 5.8

Example 5:
Work through example 5 on page 5.8-9 and record the answer below.

Multiply: $(3x - 2y)(4x + 3y)$

Example 6:
Study the solution for example 6 part a on page 5.8-10 and record your answer below. Complete parts b and c on your own and check your answers by clicking on the link. If your answers are incorrect, watch the video to find your answer.

Multiply:

a. $\left(6x^2 + 5y\right)^2$

b. $\left(4x^3 - 9y^2\right)^2$

c. $\left(2x^2y - 7)(2x^2y + 7\right)$

Example 7:
Work through example 7 on page 5.8-11 and record the answer below.

Multiply: $\left(x + 2y)(x^2 - 4xy + y^2\right)$

Read and summarize the CAUTION statement on page 5.8-11.

Section 6.1 Guided Notebook

Section 6.1 Greatest Common Factor and Factoring by Grouping

Read the list of "THINGS TO KNOW" and review any concepts you are unfamiliar with.

Section 6.1 Objective 1: Find the Greatest Common Factor of a Group of Integers

Write down the definitions for the following terms.
Factoring

Factored form

Factor (as a noun)

Factor (as a verb)

Factor over the integers

Greatest common factor

What are the three steps for **Finding the GCF of a Group of Integers**?

1.

2.

3.

Example 1:
Study the solutions for example 1 parts a and b on page 6.1-4 and record the answers below. Complete part c on your own and check your answer by clicking on the link. If your answer is incorrect, watch the video to find your error.

Find the GCF of each group of integers.

a. 36 and 60 b. 28 and 45 c. 75, 90, and 105

Section 6.1 Objective 2: Find the Greatest Common Factor of a Group of Monomials

What is the **greatest common factor (GCF) of a group of monomials**?

What are the **Common Variable Factors for a GCF**?

Example 2:
Work through example 2 and record the answers below.

Find the GCF of each group of exponential expressions.

a. x^4 and x^7

b. $y^3, y^6,$ and y^9

c. $w^6z^2, w^3z^5,$ and w^5z^4

What is the three step process for **Finding the GCF of a Group of Monomials**?

1.

2.

3.

Example 3:
Study the solutions for example 3 parts a and b on page 6.1-8 and record the answers below. Complete parts c and d on your own and check your answers by clicking on the link. If your answers are incorrect, watch the video to find your error.

Find the GCF of each group of monomials.

a. $14x^6$ and $21x^8$

b. $6a^2, 10ab,$ and $14b^2$

c. $40x^5y^6, -48x^9y,$ and $24x^2y^4$

d. $14m^3n^2, 6m^5n,$ and $9m^4$

Section 6.1 Objective 3: Factor Out the Greatest Common Factor from a Polynomial

What is the **greatest common factor (GCF) of a polynomial**?

What is the four step process for **Factoring Out the GCF from a Polynomial**?

1.

2.

3.

4.

Example 4:
Study the solution for example 4 part a on page 6.1-11 and record the answer below. Complete parts b and c on your own and check your answers by clicking on the link. If your answers are incorrect, watch the video to find your error.

Factor out the GCF from each binomial

a. $6x + 12$ b. $w^5 + w^4$ c. $8y^3 - 12y^2$

Read and summarize the CAUTION statement on page 6.1-13.

Example 5:
Study the solution for example 5 part a on page 6.1-13 and record the answer below. Complete part b on your own and check your answer by clicking on the link. If your answer is incorrect, watch the video to find your error.

Factor out the GCF from each polynomial

a. $9p^5 + 18p^4 + 54p^3$ b. $10a^4b^6 - 15a^3b^7 + 35a^2b^8$

Example 6:
Work through example 6 on page 6.1-14 and record the answer below. Check your answer by clicking on the link.

Factor out the negative sign with the GCF.

$-8x^3 + 28x^2 - 20x$

Example 7:
Study the solution for example 7 part a page 6.1-16 and record the answer below. Complete part b on your own and check your answer by clicking on the link. If your answer is incorrect, watch the video to find your error.

Factor out the common binomial factor as the GCF.

a. $4x(y + 5) + 11(y + 5)$

b. $7x(x + y) - (x + y)$

<u>**Section 6.1 Objective 4:** Factor by Grouping</u>

What is the four step process for **Factoring a Polynomial by Grouping**?

1.

2.

3.

4.

Read and summarize the CAUTION statement on page 6.1-18.

Example 8:
Study the solution for example 8 parts a and b on page 6.1-18 and record the answers below. Complete parts c and d on your own and check your answers by clicking on the link. If your answers are incorrect, watch the video to find your error.

Factor by grouping

a. $2x^2 - 6x + xy - 3y$

b. 5xy + 6 + 5x + 6y

c. $3m^2 + 3m - 2mn - 2n$

d. $4w^3 - 14w^2 - 10w + 35$

Read and summarize the CAUTION statement on page 6.1-20.

Section 6.2 Guided Notebook

Section 6.2 Factoring Trinomials of the Form $x^2 + bx + c$

Read the list of "THINGS TO KNOW" and review any concepts you are unfamiliar with.

Section 6.2 Objective 1: Factor Trinomials of the Form $x^2 + bx + c$

Work through the animation in objective 1 on page 6.2-3and take notes below.

What is the three step process for **Factoring Trinomials of the Form** $x^2 + bx + c$ **?**

1.

2.

3.

Example 1:
Study the solution for example 1 part a on page 6.2-6 and record the answer below. Complete part b on your own and check your answer by clicking on the link. If your answer is incorrect, watch the video to find your error.

Factor each trinomial.

a. $x^2 + 11x + 18$

b. $x^2 + 13x + 30$

What is a **prime number**?

What is a **prime polynomial**?

Example 2:
Work through example 2 on page 6.2-7 and record your answer below.

Factor $x^2 + 14x + 20$

Example 3:
Study the solution for example 3 part a on page 6.2-8 and record the answer below. Complete parts b and c on your own and check your answers by clicking on the link. If your answers are incorrect, watch the video to find your error.

Factor.

a. $x^2 - 13x + 40$

b. $m^2 - 5m - 36$

c. $w^2 + 7w - 60$

Section 6.2 Objective 2: Factor Trinomials of the Form $x^2 + bxy + cy^2$

Example 4:
Study the solution for example 4 part a on page 6.2-10 and record the answer below. Complete part b on your own and check your answer by clicking on the link. If your answer is incorrect, watch the video to find your error.

Factor.

a. $x^2 + 10xy + 24y^2$

b. $m^2 + 22mn - 48n^2$

Section 6.2 Objective 3: Factor Trinomials of the Form $x^2 + bx + c$ after Factoring Out the GCF

When is a polynomial **factored completely**?

Example 5:
Study the solution for example 5 part a on page 6.2-12 and record the answer below. Complete part b on your own and check your answer by clicking on the link. If your answer is incorrect, watch the video to find your error.

Factor completely.

a. $4x^2 - 28x - 32$

b. $2y^3 - 36y^2 + 64y$

Example 6:
Work through example 6 on page 6.2-14 and record the answer below. Check your answer by clicking on the link. If your answer is incorrect, watch the video to find your error.

Factor $-x^2 + 3x + 10$

Section 6.3 Guided Notebook

Section 6.3 Factoring Trinomials of the Form $ax^2 + bx + c$ Using Trial and Error

Read the list of "THINGS TO KNOW" and review any concepts you are unfamiliar with.

Section 6.3 Objective 1: Factor Trinomials of the Form $ax^2 + bx + c$ Using Trial and Error

Work through the animation in objective 1 on page 6.3-3 and take notes below.

What is the four step strategy for **Factoring Trinomials of the Form $ax^2 + bx + c$?**

1.

2.

3.

4.

Example 1:
Work through example 1 on page 6.3-5 and record the answer below. Check your answer by clicking on the link.

Factor $3x^2 + 7x + 2$

Read and summarize the CAUTION on page 6.3-6.

Example 2:
Work through example 2 on page 6.3-6 and record the answer below. Check your answer by clicking on the link. If your answer is incorrect, watch the video to find your error.

Factor $5x^2 + 17x + 6$

Example 3:
Work through example 3 on page 6.3-8 and record the answers below. Check your answers by clicking on the link. If your answers are incorrect, watch the video to find your error.

Factor

a. $4x^2 - 5x - 6$

b. $12n^2 - 16n + 5$

Read and summarize the CAUTION statement on page 6.3-9.

Example 4:
Work through example 4 on page 6.3-10 and record the answer below.

Factor $2y^2 - 19y + 15$

Section 6.3 Objective 2: Factor Trinomials of the Form $ax^2 + bxy + cy^2$ Using Trial and Error

Example 5:
Study the solution for example 5 part a on page 6.3-12 and record the answer below. Complete part b on your own and check your answer by clicking on the link. If your answer is incorrect, watch the video to find your error.

Factor.

a. $6x^2 + 17xy - 3y^2$

b. $2m^2 + 11mn + 12n^2$

Section 6.4 Guided Notebook

Section 6.4 Factoring Trinomials of the Form $ax^2 + bx + c$ Using the *ac* Method

Read the list of "THINGS TO KNOW" and review any concepts you are unfamiliar with.

Section 6.4 Objective 1: Factor Trinomials of the Form $ax^2 + bx + c$ Using the *ac* Method

What are the five steps for the ***ac* Method for Factoring Trinomials of the Form $ax^2 + bx + c$**

1.

2.

3.

4.

5.

Work through the animation in objective 1, page 6.4-3 and take notes below.

What are the two other names for the *ac* Method?

Example 1:
Work through example 1 on page 6.4-4 and record the answer below. Check your answer by clicking on the link. If your answer is incorrect, watch the video to find your error.

Factor $3x^2 + 14x + 8$ using the *ac* method.

Example 2:
Work through example 2 on page 6.4-6 and record the answer below. Check your answer by clicking on the link. If your answer is incorrect, watch the video to find your error.

Factor $2x^2 - 3x - 20$ using the *ac* method.

Example 3:
Work through example 3 on page 6.4-7 and record the answers below. Check your answers by clicking on the link. If your answers are incorrect, watch the video to find your error.

Factor each trinomial using the *ac* method. If the trinomial is prime, state this as your answer.

a. $2x^2 + 9x - 18$

b. $6x^2 - 23x + 20$

c. $5x^2 + x + 6$

Section 6.4 Objective 2: Factor Trinomials of the Form $ax^2 + bxy + cy^2$ Using the *ac* Method

Example 4:
Work through example 4 on page 6.4-8 and record the answer below. Check your answer by clicking on the link. If your answer is incorrect, watch the video to find your error.

Factor $2p^2 + 7pq - 15q^2$ using the *ac* method.

Section 6.4 Objective 2: Factor Trinomials of the Form $ax^2 + bx + c$ Using the ac Method after Factoring out the GCF

Example 5:
Work through example 5 on page 6.4-10 and record the answer below. Check your answer by clicking on the link. If your answer is incorrect, watch the video to find your error.

Factor completely: $24t^5 - 52t^4 - 20t^3$

Example 6:
Work through example 6 on page 6.4-11 and record the answer below. Check your answer by clicking on the link. If your answer is incorrect, watch the video to find your error.

Factor completely: $-2x^2 + 9x + 35$

Section 6.5 Guided Notebook

Section 6.5 Factoring Special Forms

Read the list of "THINGS TO KNOW" and review any concepts you are unfamiliar with.

Section 6.5 Objective 1: Factor the Difference of Two Squares

Summarize **Factoring the Difference of Two Squares**.

Read and summarize the CAUTION statement on page 6.5-4.

Example 1:
Work through example 1 on page 6.5-4 and record the answers below.

Factor each expression completely.

a. $x^2 - 9$

b. $16 - y^2$

What is a **perfect square**? Give two examples of a perfect square.

Example 2:
Study the solution for example 2 part a on page 6.5-6 and record the answer below. Complete parts b–d on your own and check your answers by clicking on the link. If your answers are incorrect, watch the video to find your error.

Factor each expression completely.

a. $z^2 - \frac{25}{16}$

b. $36x^2 - 25$

c. $4 - 49n^6$

d. $81m^2 - n^2$

Example 3:
Study the solution for example 3 part a on page 6.5-7 and record the answer below. Complete part b on your own and check your answer by clicking on the link. If your answer is incorrect, watch the video to find your error.

Factor each expression completely.

a. $3x^2 - 75$

b. $36x^3 - 64x$

What does the CAUTION statement on page 6.5-8 say about the *sum of two perfect squares*?

Example 4:
Work through example 4 on page 6.5-9 and record the answer below.

Factor completely.

$16x^4 - 81$

Section 6.5 Objective 2: Factor Perfect Square Trinomials

Summarize **Factoring Perfect Square Trinomials.**

Example 5:
Work through example 5 on page 6.5-11 and record the answers below

Factor each expression completely.

a. $x^2 + 6x + 9$

b. $y^2 - 10y + 25$

Example 6:
Study the solution for example 6 part a on page 6.5-12 and record the answer below. Complete part b on your own and check your answer by clicking on the link. If your answer is incorrect, watch the video to find your error.

Factor each expression completely.

a. $4x^2 + 12x + 9$

b. $25y^2 - 60y + 36$

Example 7:
Work through example 7 on page 6.5-13 and record the answers below. Check your answers by clicking on the link. If your answers are incorrect, watch the video to find your error.

Factor each expression completely.

a. $16x^2 + 24xy + 9y^2$

b. $m^4 - 12m^2 + 36$

Section 6.5 Objective 3: Factor the Sum or Difference of Two Cubes

Summarize **Factoring the Sum and Difference of Two Cubes**

What are **perfect cubes**? Give two examples of a perfect cube.

Example 8:
Work through example 8 on page 6.5-16 and record the answers below.

Factor each expression completely.

a. $x^3 + 64$

b. $z^3 - 8$

Example 9:
Study the solution for example 9 part a on page 6.5-18 and record the answer below. Complete parts b and c on your own and check your answers by clicking on the link. If your answers are incorrect, watch the video to find your error.

Factor each expression completely.

a, $125y^3 - 1$

b. $128z^3 + 54y^3$

c. $8x^3y^3 + y^5$

Section 6.6 Guided Notebook

Section 6.6 A General Factoring Strategy

Read the list of "THINGS TO KNOW" and review any concepts you are unfamiliar with.

Section 6.6 Objective 1: Factor Polynomials Completely

What is the four-step **General Strategy for Factoring Polynomials Completely**?

1.

2.

 a.

 b.

 c.

3.

4.

Example 1:
Study the solutions for example 1 parts a and b on page 6.6-4 and record the answers below. Complete parts c and d on your own and check your answers by clicking on the link. If your answers are incorrect, watch the video to find your error.

Factor each expression completely.

a. $w^2 - w - 20$

b. $4y^4 - 32y$

c. $x^2 - 14x + 49$

d. $3z^3 - 15z^2 - 42z$

Example 2:
Study the solution for example 2 part a on page 6.6-7 and record the answer below. Complete parts b and c on your own and check your answers by clicking on the link. If your answers are incorrect, watch the video to find your error.

Factor each expression completely.

a. $2x^3 - 5x^2 - 8x + 20$

b. $3a^2 - 10a - 8$

c. $3z^2 + z - 1$

Example 3:
Work through example 3 part a on page 6.6-8 and record the answer below. Complete parts b–d on your own and check your answers by clicking on the link. If your answers are incorrect, watch the video to find your error.

Factor each expression completely.

a. $10x^2 + 11xy - 6y^2$

b. $2p^2 - 32pq + 128q^2$

c. $7x^2z - 14x$

d. $-3y^4z - 24yz^4$

Section 6.7 Guided Notebook

Section 6.7 Solving Polynomial Equations by Factoring

Read the list of "THINGS TO KNOW" and review any concepts you are unfamiliar with.

Section 6.7 Objective 1: Solve Quadratic Equations by Factoring

Write down the definitions for the following terms.

Polynomial equation

Standard form

Degree of a polynomial equation

Quadratic equations

What is the **Zero Product Property**?

Read and summarize the CAUTION statement on page 6.7-4.

Example 1:
Study the solution for example 1 part a on page 6.7-5 and record the answer below. Complete part b on your own and check your answer by clicking on the link. If your answer is incorrect, watch the video to find your error.

Solve each equation.

a. $(x + 10)(x - 3) = 0$

b. $x(3x + 5) = 0$

What is the four-step process for **Solving Polynomial Equations by Factoring**?

1.

2.

3.

4.

Example 2:
Study the solution for example 2 part a on page 6.7-6 and record the answer below. Complete part b on your own and check your answer by clicking on the link. If your answer is incorrect, watch the video to find your error.

Solve each equation by factoring.

a. $z^2 + 4z - 12 = 0$

b. $-4x^2 + 28x - 40 = 0$

Read and summarize the CAUTION statement on page 6.7-8.

Example 3:
Study the solution for example 3 part a on page 6.7-8 and record the answer below. Complete parts b and c on your own and check your answers by clicking on the link. If your answers are incorrect, watch the video to find your error.

Solve each equation by factoring.

a. $9w^2 + 64 = 48w$

b. $4m^2 = 49$

c. $3x(x - 2) = 2 - x$

Section 6.7

Read and summarize the CAUTION statement on page 6.6-10.

Example 4:
Work through example 4 on page 6.7-11 and record the answers below. Check your answers by clicking on the link. If your answers are incorrect, watch the video to find your error.

Solve each equation by factoring.

a. $(x + 2)(x - 5) = 18$

b. $(x + 3)(3x - 5) = 5(x + 1) - 10$

<u>**Section 6.7 Objective 2:** Solve Polynomial Equations by Factoring</u>

Example 5:
Study the solution for example 5 part a on page 6.7-13 and record the answer below. Complete parts b and c on your own and check your answers by clicking on the link. If your answers are incorrect, watch the video to find your error.

a. $(x + 7)(2x - 1)(5x + 4) = 0$

b. $24x^3 + 8x^2 = 100x^2 - 28x$

c. $z^3 + z^2 = z + 1$

Example 6:
Work through example 6 on page 6.7-15 and record the answer below.

Solve by factoring.

$(2x - 9)(3x^2 - 16x - 12) = 0$

Section 6.8 Guided Notebook

Section 6.8 Applications of Quadratic Equations

Read the list of "THINGS TO KNOW" and review any concepts you are unfamiliar with.

Section 6.8 Objective 1: Solve Application Problems Involving Consecutive Numbers

Read and summarize the CAUTION statement on page 6.8-3.

Example 1:
Work through example 1 on page 6.8-3 and record the answer below.

The house numbers on the west side of a street are consecutive positive odd integers. The product of the house numbers for two next-door-neighbors on the west side of the street is 575. Find the house numbers.

Section 6.8 Objective 2: Solve Application Problems Involving Geometric Figures

Example 2:
Work through example 2 on page 6.8-6 and record the answer below. Check your answer by clicking on the link. If your answer is incorrect, watch the video to find your error.

A swimming pool is 20 feet wide and 30 feet long. A sidewalk border around the pool has uniform width and an area that is equal to the area of the pool. Find the width of the border.

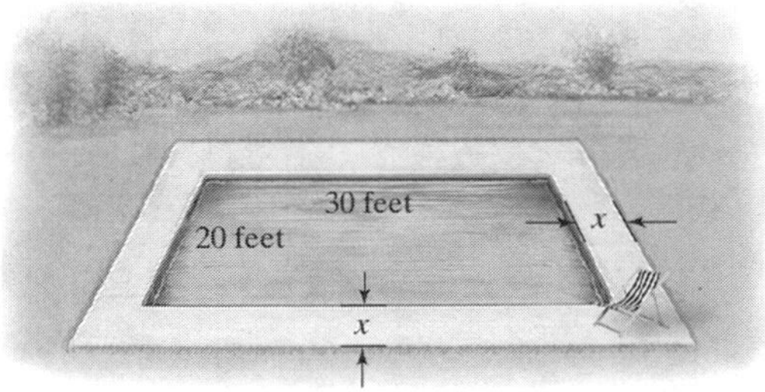

Section 6.8 Objective 3: Solve Application Problems Using the Pythagorean Theorem

Write down the definition for the following terms.

Right triangles

Right angle

Hypotenuse

Legs

What is the **Pythagorean Theorem**?

Example 3:
Work through example 3 on page 6.8-8 and record the answer below. Check your answer by clicking on the link. If your answer is incorrect, watch the video to find your error.

A wire is attached to a cell phone tower for support. The length of the wire is 40 meters less than twice the height of the tower. The wire is fixed to the ground at a distance that is 40 meters more than the height of the tower. Find the length of the wire.

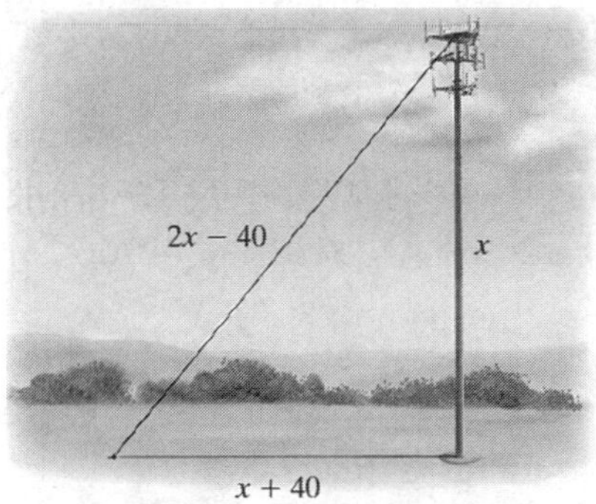

Read and summarize the CAUTION statement on page 6.8-9.

Section 6.8 Objective 4: Solve Application Problems Involving Quadratic Models

Example 4:
Work through example 4 on page 6.8-10 and record the answer below. Watch the video for a detailed solution.

The Grand Canyon Skywalk sits 4000 ft above the Colorado River. If an object is dropped from the observation deck, its height h, in feet after t seconds, is given by

$$h = -16t^2 + 4000.$$

How long will it take for the object to be 400 feet above the Colorado River?

Example 5:
Work through example 5 on page 6.8-12 and record the answer below. Check your answer by clicking on the link. If your answer is incorrect, watch the video to find your error.

For household incomes under \$100,000, the relationship between the percentage of households with home broadband access and the annual household income can be approximated by the model,

$$y = -0.01x^2 + 1.7x + 9.5.$$

Here, x is the annual household income (in \$1000s) and y is the percentage of households with home broadband access. Use the model to estimate the annual household income if 75.5 percent of such households have home broadband access.

Section 7.1 Guided Notebook

Section 7.1 Simplifying Rational Expressions

Read the list of "THINGS TO KNOW" and review any concepts you are unfamiliar with.

Section 7.1 Objective 1: Evaluate Rational Expressions

Write down the definition for a **Rational Expression** and give two examples.

Example 1:

Work through example 1 on page 7.1-4 and record the answers below.

Evaluate $\frac{x+8}{x-2}$ for the given value of x.

a. $x = 4$

b. $x = -6$

Example 2:

Work through example 2 on page 7.1-5 and record the answer below. Check your answer by clicking on the link. If your answer is incorrect, watch the video to find your error.

Evaluate $\frac{x^2 - y}{9x+5y}$ for $x = 3$ and $y = -1$.

Section 7.1 Objective 2: Find Restricted Values for Rational Expressions

Write down the definition of a **Restricted Value**.

What is the technique for **Finding Restricted Values for Rational Expressions in One Variable**?

Example 3:
Study the solution for example 3 part a on page 7.1-7 and record the answer below. Complete part b on your own and check your answer by clicking on the link. If your answer is incorrect, watch the video to find your error.

Find any restricted values for each rational expression.

a. $\dfrac{3x+5}{3x-2}$

b. $\dfrac{x^2+2x-35}{x^2+x-30}$

Example 4:
Work through example 4 on page 7.1-8 and record the answers below.

Find any restricted values for each rational expression.

a. $\dfrac{2x+9}{4}$

b. $\dfrac{2x}{x^2+1}$

Section 7.1 Objective 3: Simplify Rational Expressions

When is a fraction written in **lowest terms** or **simplest form**?

What is the **Simplification Principle for Rational Expressions**?

What are the three steps for **Simplifying Rational Expressions**?

1.

2

3.

Example 5:
Work through example 5 on page 7.1-11 and record the answer below.

Simplify $\dfrac{2x^2-6x}{7x-21}$

Example 6:
Work through example 6 on page 7.1-12 and record the answer below.

Simplify $\dfrac{5x}{x^2+5x}$

Read and summarize the CAUTION statements on pages 7.1-12 and 7.1-13.

Example 7:
Work through example 7 on page 7.1-13 and record the answer below. Check your answer by clicking on the link. If your answer is incorrect, watch the video to find your error.

Simplify $\dfrac{y^2+2y-24}{y^2+4y-32}$

Example 8:
Work through example 8 on page 7.1-14 and record the answer below. Check your answer by clicking on the link. If your answer is incorrect, watch the video to find your error.

Simplify $\dfrac{2m^2+m-15}{2m^3-5m^2-18m+45}$

Example 9:
Work through example 9 on page 7.1-14 and record the answer below. Check your answer by clicking on the link. If your answer is incorrect, watch the video to find your error.

Simplify $\dfrac{x^2-xy-12y^2}{2x^2+7xy+3y^2}$

Example 10:
Work through example 10 on page 7.1-15 and record the answer below.

Simplify $\dfrac{w^2 - y^2}{2xy + 2xw}$

Example 11:
Work through example 11 on page 7.1-16 and record the answer below.

Simplify $\dfrac{3x - 10}{10 - 3x}$

Example 12:
Work through example 12 on page 7.1-17 and record the answer below. Check your answer by clicking on the link. If your answer is incorrect, watch the video to find your error.

Simplify $\dfrac{2x^2 - 27x + 70}{49 - 4x^2}$

Section 7.2 Guided Notebook

Section 7.2 Multiplying and Dividing Rational Expressions

Read the list of "THINGS TO KNOW" and review any concepts you are unfamiliar with.

Section 7.2 Objective 1: Multiply Rational Expressions

What are the three steps for **Multiplying Rational Expressions**?

1.

2.

3.

Read and summarize the CAUTION statement on page 7.2-4.

Example 1:
Work through example 1 on page 7.2-5 and record the answer below.

Multiply $\frac{5x^2}{2y} \cdot \frac{6y^2}{25x^3}$

Example 2:
Work through example 2 on page 7.2-6 and record the answer below.

Multiply $\frac{3x-6}{2x} \cdot \frac{8}{5x-10}$

Example 3:
Work through example 3 on page 7.2-7 and record the answer below. Check your answer by clicking on the link. If your answer is incorrect, watch the video to find your error.

Multiply $\frac{x^2-4}{x^2+2x-35}\cdot\frac{x^2-25}{x+2}$

Example 4:
Work through example 4 on page 7.2-8 and record the answer below. Watch the video for a detailed solution.

Multiply $\frac{2x^2+3x-2}{3x^2-2x-1}\cdot\frac{3x^2+4x+1}{2x^2+x-1}$

Example 5:
Work through example 5 on page 7.2-8 and record the answer below.

Multiply $\frac{3x^2+9x+27}{x-1}\cdot\frac{x+3}{x^3-27}$

Example 6:
Work through example 6 on page 7.2-10 and record the answer below. Check your answer by clicking on the link. If your answer is incorrect, watch the video to find your error.

Multiply $\dfrac{3x^2+10x-8}{2x-3x^2}\cdot\dfrac{4x+1}{x+4}$

Example 7:
Work through example 7 on page 7.2-11 and record the answer below. Check your answer by clicking on the link. If your answer is incorrect, watch the video to find your error.

Multiply $\dfrac{x^2+xy}{3x+y}\cdot\dfrac{3x^2+7xy+2y^2}{x^2-y^2}$

Section 7.2 Objective 2: Divide Rational Expressions

What is the two step process for **Dividing Rational Expressions**?

1.

2.

Example 8:
Work through example 8 on page 7.2-13 and record the answers below.

Divide each rational expression.

a. $\dfrac{6x^5}{9y^3} \div \dfrac{5x^4}{3y^2}$

b. $\dfrac{(x+2)(x-1)}{(3x-5)} \div \dfrac{(x-1)(x+4)}{(2x+3)}$

Example 9:
Work through example 9 on page 7.2-14 and record the answer below. Check your answer by clicking on the link. If your answer is incorrect, watch the video to find your error.

Divide $\dfrac{9y^2-81}{4y^2} \div \dfrac{y+3}{8}$

Example 10:
Work through example 10 on page 7.2-16 and record the answer below. Check your answer by clicking on the link. If your answer is incorrect, watch the video to find your error.

Divide $\dfrac{2x^2+21x+40}{3x^2+23x-8} \div \dfrac{4x^2+16x+15}{x+2}$

Example 11:
Work through example 11 on page 7.2-16 and record the answer below. Check your answer by clicking on the link. If your answer is incorrect, watch the video to find your error.

Divide $\dfrac{x^3-8}{2x^2-x-6} \div \dfrac{x^2+2x+4}{6x^2+11x+3}$

Example 12:
Work through example 12 on page 7.2-17 and record the answer below. Check your answer by clicking on the link. If your answer is incorrect, watch the video to find your error.

Divide $\dfrac{x^3-8y^3}{3x+y} \div \dfrac{4x-8y}{6x^2+17xy+5y^2}$

Example 13:
Work through example 13 on page 7.2-18 and record the answer below. Check your answer by clicking on the link. If your answer is incorrect, watch the video to find your error.

Perform the indicated operations.

$$\frac{x^2+2x-15}{x^2+2x-8} \cdot \frac{x^2+3x+2}{x^2+4x-21} \div \frac{x+2}{x^2+9x+14}$$

Section 7.3 Guided Notebook

Section 7.3 Least Common Denominators

Read the list of "THINGS TO KNOW" and review any concepts you are unfamiliar with.

Section 7.3 Objective 1: Find the Least Common Denominator of Rational Expressions

What are the three steps for **Finding the Least Common Denominator (LCD) of Rational Expressions**?

1.

2.

3.

Read and summarize the CAUTION statement on page 7.3-4.

Example 1:
Work through example 1 on page 7.3-4 and record the answers below.

Find the LCD of the rational expressions.

a. $\frac{7}{10x^3}, \frac{3}{5x^2}$

b. $\frac{x+2}{3x}, \frac{x-1}{2x^2+6x}$

Example 2:
Work through example 2 on page 7.3-6 and record the answers below. Check your answers by clicking on the link. If your answers are incorrect, watch the video to find your error.

Find the LCD of the rational expressions.

a. $\dfrac{z^2}{6-z}, \dfrac{9}{2z-12}$

b. $\dfrac{y+2}{y^2+2y-3}, \dfrac{2y}{y^2+5y+6}$

Example 3:
Work through example 3 on page 7.3-7 and record the answers below. Check your answers by clicking on the link. If your answers are incorrect, watch the video to find your error.

Find the LCD of the rational expressions.

a. $\dfrac{4x}{10x^2-7x-12}, \dfrac{2x-3}{5x^2-11x-12}$

b. $\dfrac{10-x}{6x^2+5x+1}, \dfrac{-4}{9x^2+6x+1}, \dfrac{x^2-7x}{10x^2-x-3}$

<u>**Section 7.3 Objective 2:** Write Equivalent Rational Expressions</u>

When writing equivalent fractions, what is the key question that should be asked?

What is the three step process for **Writing Equivalent Rational Expressions**?

1.

2.

3.

Section 7.3

Read and summarize the CAUTION statement on page 7.3-9.

Example 4:
Work through example 4 on page 7.3-10 and record the answers below.

Write each rational expression as an equivalent rational expression with the desired denominator.

a. $\frac{3}{2x} = \frac{\quad}{10x^3}$

b. $\frac{x+2}{3x+15} = \frac{\quad}{3(x-1)(x+5)}$

Example 5:
Work through example 5 on page 7.3-11 and record the answers below. Check your answers by clicking on the link. If your answers are incorrect, watch the video to find your error.

Write each rational expression as an equivalent rational expression with the desired denominator..

a. $\frac{-7}{1-4y} = \frac{\quad}{8y^2-2y}$

b. $\frac{5z}{z^2+z-6} = \frac{\quad}{(z-4)(z-2)(z+3)}$

Section 7.4 Guided Notebook

Section 7.4 Adding and Subtracting Rational Expressions

Read the list of "THINGS TO KNOW" and review any concepts you are unfamiliar with.

Section 7.4 Objective 1: Add and Subtract Rational Expressions with Common Denominators

Write the technique for **Adding and Subtracting Rational Expressions with Common Denominators**.

Example 1:
Work through example 1 on page 7.4-4 and record the answers below.

Add or subtract.

a. $\frac{4z}{3}+\frac{5z}{3}$

b. $\frac{3r}{7s^2}-\frac{2r}{7s^2}$

Read and summarize the CAUTION statement on page 7.4-5.

Example 2:
Work through example 2 on page 7.4-6 and record the answers below. Check your answers by clicking on the link. If your answers are incorrect, watch the video to find your error.

Add or subtract.

a. $\frac{9x}{x-4}+\frac{7x-2}{x-4}$

b. $\frac{5y+1}{y-2}-\frac{2y+3}{y-2}$

Example 3:
Work through example 3 on page 7.4-8 and record the answers below. Check your answers by clicking on the link. If your answers are incorrect, watch the video to find your error.

Add or subtract.

a. $\dfrac{4}{x^2+2x-8}+\dfrac{x}{x^2+2x-8}$ b. $\dfrac{x}{x+2}-\dfrac{x-3}{x+2}$ c. $\dfrac{x^2-2}{x-5}-\dfrac{4x+3}{x-5}$

<u>**Section 7.4 Objective 2:** Add and Subtract Rational Expressions with Unlike Denominators</u>

What is the four step process for **Adding and Subtracting Rational Expressions with Unlike Denominators**?

1.

2.

3.

4.

Example 4:
Study the solution for example 4 part a on 7.4-10 and record the answer below. Complete part b on your own and check your answer by clicking on the link. If your answer is incorrect, watch the video to find your error.

Perform the indicated operations and simplify.

a. $\dfrac{7}{6x}+\dfrac{3}{2x^3}$ b. $\dfrac{3x}{x-3}-\dfrac{x-2}{x+3}$

Example 5:
Study the solution for example 5 part a on 7.4-13 and record the answer below. Complete part b on your own and check your answer by clicking on the link. If your answer is incorrect, watch the video to find your error.

Perform the indicated operations and simplify.

a. $\frac{z+2}{3z}-\frac{5}{3z+12}$ b. $\frac{5}{4m-12}+\frac{3}{2m}$

Example 6:
Study the solution for example 6 part a on 7.4-15 and record the answer below. Complete part b on your own and check your answer by clicking on the link. If your answer is incorrect, watch the video to find your error.

Perform the indicated operations and simplify.

a. $2+\frac{4}{x-5}$ b. $\frac{x^2-2}{x^2+6x+8}-\frac{x-3}{x+4}$

Example 7:
Work through example 7 on page 7.4-18 and record the answers below. Check your answers by clicking on the link. If your answers are incorrect, watch the video to find your error.

Perform the indicated operations and simplify.

a. $\frac{x+7}{x^2-9}+\frac{3}{x+3}$

b. $\frac{x+1}{2x^2+5x-3}-\frac{x}{2x^2+3x-2}$

Example 8:
Work through example 8 on page 7.4-19 and record the answer below.

Perform the indicated operations and simplify.

$\frac{2y}{y-5}+\frac{y-1}{5-y}$

Example 9:
Work through example 9 on page 7.4-20 and record the answer below. Check your answer by clicking on the link. If your answer is incorrect, watch the video to find your error.

$\frac{x+1}{x^2-6x+9}+\frac{3}{x-3}-\frac{6}{x^2-9}$

Section 7.5 Guided Notebook

Section 7.5 Complex Rational Expressions

Read the list of "THINGS TO KNOW" and review any concepts you are unfamiliar with.

Section 7.5 Objective 1: Simplify Complex Rational Expressions by First Simplifying the Numerator and Denominator

Write down the definition of a **Complex Rational Expression**. Provide one example.

What form is a **simplified complex rational expression** written in?

To simplify a complex fraction using Method I divide the ____________________ by the ____________________.

Example 1:
Work through example 1 on page 7.5-4 and record the answer below.

Simplify $\dfrac{\dfrac{2}{9x}}{\dfrac{5}{6xy}}$

What are the three steps for **Method I for Simplifying Complex Rational Expressions**?

1.

2.

3.

Example 2:
Study the solution for example 2 part a on page 7.5-6 and record the answer below. Complete part b on your own and check your answer by clicking on the link. If your answer is incorrect, watch the video to find your error.

Use Method I to simplify each complex rational expression.

a. $\dfrac{\frac{1}{3}-\frac{1}{x}}{\frac{1}{9}-\frac{1}{x^2}}$

b. $\dfrac{4-\frac{5}{x-1}}{\frac{6}{x-1}-7}$

Section 7.5 Objective 2: Simplify Complex Rational Expressions by Multiplying by a Common Denominator

Work through example 3 on page 7.5-9 and record the answer below.

Simplify $\dfrac{\frac{2}{9x}}{\frac{5}{6xy}}$

What is the three step process for **Method II for Simplifying Complex Rational Expressions**

1.

2.

3.

Example 4:
Study the solution for example 4 part a on page 7.5-11 and record the answer below. Complete part b on your own and check your answer by clicking on the link. If your answer is incorrect, watch the video to find your error.

Use Method II to simplify each complex rational expression.

a. $\dfrac{\frac{1}{3}-\frac{1}{x}}{\frac{1}{9}-\frac{1}{x^2}}$

b. $\dfrac{4-\frac{5}{x-1}}{\frac{6}{x-1}-7}$

Which of the two methods do you prefer? Why?

Example 5:
Work through example 5 on page 7.5-14 and record the answer below. Check your answer by clicking on the link. If your answer is incorrect, watch the video to find your error.

Simplify the complex rational expressions using Method I or Method II.

$$\frac{\frac{5}{n-2}-\frac{3}{n}}{\frac{6}{n^2-2n}+\frac{2}{n}}$$

Example 6:
Work through example 6 on page 7.5-15 and record the answer below. Check your answer by clicking on the link. If your answer is incorrect, watch the video to find your error.

Simplify the complex rational expression.

$$\frac{1-9y^{-1}+14y^{-2}}{1+3y^{-1}-10y^{-2}}$$

Section 7.6 Guided Notebook

Section 7.6 Solving Rational Equations

Read the list of "THINGS TO KNOW" and review any concepts you are unfamiliar with.

Section 7.6 Objective 1: Identify Rational Equations

What is the definition of a **Rational Equation**?

Example 1:
Work through example 1 on page 7.6-3 and record the answers below.

Determine if each statement is a rational equation. If not, state why.

a. $\frac{x-4}{x}+\frac{4}{x+5}=\frac{6}{x}$

b. $\frac{5}{y}+\frac{7}{y+2}$

c. $\frac{\sqrt{k+1}}{k+3}=\frac{k-5}{k+4}$

d. $5n^{-1}=3n^{-2}$

Section 7.6 Objective 2: Solve Rational Equations

Example 2:
Study the solution for example 2 part a on page 7.6-5 and record the answer below. Complete part b on your own and check your answer by clicking on the link. If your answer is incorrect, watch the video to find your error.

Solve.

a. $\frac{1}{2}x+\frac{2}{3}=\frac{3}{4}$

b. $\frac{1}{x}+\frac{1}{2}=\frac{1}{3}$

Section 7.6

What are **extraneous solutions**?

What is the five step process for **Solving Rational Equations**?

1.

2.

3.

4.

5.

Example 3:
Work through example 3 on page 7.6-8 and record the answer below.

Solve $\frac{2}{x}-\frac{x-3}{2x}=3$

Example 4:
Work through example 4 on page 7.6-10 and record the answer below. Check your answer by clicking on the link. If your answer is incorrect, watch the video to find your error.

Solve $\frac{4}{5}-\frac{3}{x-3}=\frac{1}{x}$

Example 5:
Work through example 5 on page 7.6-11 and record the answer below. Check your answer by clicking on the link. If your answer is incorrect, watch the video to find your error.

Solve $\frac{m}{m+2}+\frac{5}{m-2}=\frac{20}{m^2-4}$

Example 6:
Work through example 6 on page 7.6-13 and record the answer below. Check your answer by clicking on the link. If your answer is incorrect, watch the video to find your error.

Solve $\frac{2}{x-3}-\frac{4}{x^2-2x-3}=\frac{1}{x+1}$

<u>Section 7.6 Objective 3:</u> <u>Identify and Solve Proportions</u>

Write down the definition for the following terms.

Ratio

Proportion

Cross-multiplying

Example 7:
Study the solution for example 7 part a on page 7.6-15 and record the answer below. Complete part b on your own and check your answer by clicking on the link. If your answer is incorrect, watch the video to find your error.

Solve.

a. $\dfrac{8}{x+3}=\dfrac{5}{x}$

b. $\dfrac{x}{6}=\dfrac{2}{x-1}$

Read and summarize the CAUTION statement on page 7.6-16.

Section 7.6 Objective 4: Solve a Formula Containing Rational Expressions for a Given Variable

What is a **formula**?

Example 8:
Study the solution for example 8 part a on page 7.6-17 and record the answer below. Complete part b on your own. Watch the video to see the complete solution.

Solve each formula for the given variable.

a. $I=\dfrac{E}{r+R}$ for R

b. $\dfrac{1}{f}=\dfrac{1}{c}+\dfrac{1}{d}$ for d

Read and summarize the CAUTION statement on page 7.6-18.

Section 7.7 Guided Notebook

Section 7.7 Applications of Rational Equations

Read the list of "THINGS TO KNOW" and review any concepts you are unfamiliar with.

Section 7.7 Objective 1: Use Proportions to Solve Problems

What is a **proportion**?

Example 1:
Work through example 1 on page 7.7-3 and record the answer below.

A quality-control inspector examined a sample of 200 light bulbs and found 18 of them to be defective. At this ratio, how many defective bulbs can the inspector expect in a shipment of 22,000 light bulbs?

Example 2:
Work through example 2 on page 7.7-5 and record the answer below. Check your answer by clicking on the link. If your answer is incorrect, watch the video to find your error.

A landscaper plants grass seed at a general rate of 7 pounds for every 1000 square feet. If the landscaper has 25 pounds of grass seed on hand, how many additional pounds of grass seed will he need to purchase for a job to plant grass o a 45,000 square-foot yard?

What are **similar triangles**?

In similar triangles corresponding angles are ________________________________.

In similar triangles corresponding sides are ________________________________.

Example 3:
Work through example 3 on page 7.7-6 and record the answer below.

Find the unknown length for n for the following similar triangles.

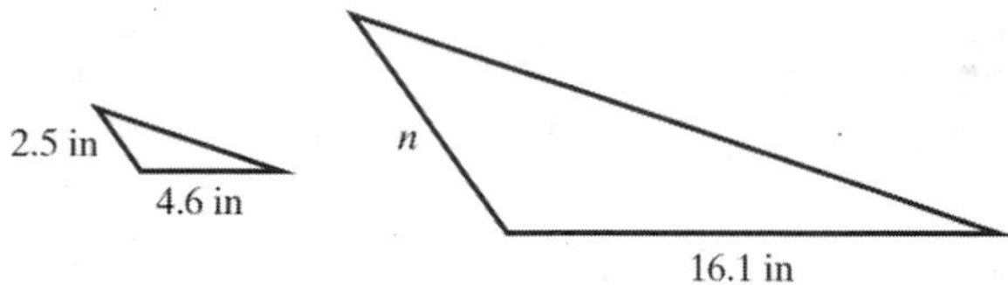

Example 4:
Work through example 4 on page 7.7-8 and record the answer below. Check your answer by clicking on the link. If your answer is incorrect, watch the video to find your error.

A forest ranger wants to determine the height of a tree. She measures the tree's shadow as 84 feet long. Her own shadow at the same time is 7.5 feet long. If she is 5.5 feet tall, how tall is the tree?

<u>**Section 7.7 Objective 2:** Use Formulas Containing Rational Expressions to Solve Problems</u>

Example 5:
Work through example 5 on page 7.7-9 and record the answer below.

In electronics, the total resistance R of a circuit containing two resistors in parallel is given by the formula $\frac{1}{R}=\frac{1}{R_1}+\frac{1}{R_2}$, where R_1 and R_2 are the two individual resistances. If the total resistance is 10 ohms and one resistor has twice the resistance of the other, find the resistance of each circuit.

Section 7.7 Objective 3: Solve Uniform Motion Problems Involving Rational Equations

Example 6:
Work through example 6 on page 7.7-11 and record the answer below. Check your answer by clicking on the link. If your answer is incorrect, watch the video to find your error.

Emalie can travel 16 miles upriver in the same amount of time it takes her to travel 24 miles downriver. If the speed of the current is 4 mph, how fast can her boat travel in still water?

Example 7:
Work through example 7 on page 7.7-13 and record the answer below. Check your answer by clicking on the link. If your answer is incorrect, watch the video to find your error.

Fatima rode an express train 223.6 miles from Boston to New York City and then rode a passenger train 218.4 miles from New York City to Washington, D.C. If the express train travels 30 miles per hour faster than the passenger train and her total trip took 6.5 hours, what was the average speed of the express train?

Section 7.7 Objective 4: Solve Problems Involving Rate of Work

What is a **rate of work**?

Can rates be added?

Can times be added?

Example 8:
Work through example 8 on page 7.7-16 and record the answer below.

Avril can paint a room in 4 hours if she works alone. Anisa can paint the same room in 2 hours if she works alone. How long will it take the two women to paint the room if they work togeher?

Example 9:
Work through example 9 on page 7.7-18 and record the answer below. Check your answer by clicking on the link. If your answer is incorrect, watch the video to find your error.

A small pump takes 8 more hours than a larger pump to empty a pool. Together, the pumps can empty the pool in 3 hours. How long will it take the larger pump to empty the pool if it works alone?

Example 10:
Work through Example 10 on page 7.7-19 and record the answer below. Watch the animation to see the complete solution.

A garden hose can fill a pond in 2 hours whereas an outlet pipe can drain the pond in 10 hours. If the outlet pipe is accidentally left open, how long would it take to fill the pond?

Section 7.8 Guided Notebook

Section 7.8 Variation

Read the list of "THINGS TO KNOW" and review any concepts you are unfamiliar with.

Section 7.8 Objective 1: Solve Problems Involving Direct Variation

What is the purpose of **direct variation**?

Write down the definition of **Direct Variation**.

Example 1:
Work through example 1 on page 7.8-3 and record the answers below.

Suppose y varies directly with x, and $y = 20$ when $x = 8$.

a. Find the equation that relates x and y.

b. Find y when $x = 12$.

Example 2:
Work through example 2 on page 7.8-5 and record the answers below. Check your answers by clicking on the link. If your answers are incorrect, watch the video to find your error.

Suppose y varies directly with the cube of x, and $y = 375$ when $x = 5$.

a. Find the equation that relates x and y.

b. Find y when $x = 2$.

Section 7.8

What is the four step process for **Solving Variation Problems**?

1.

2.

3.

4.

Example 3:
Work through example 3 on page 7.8-6 and record the answer below.

The kinetic energy of an object in motion varies directly with the square of its speed. If a van traveling at the speed of 30 meters per second has 945,000 joules of kinetic energy, how much kinetic energy does it have it if is traveling at a speed of 20 meters per second?

Example 4:
Work through example 4 on page 7.8-7 and record the answer below. Check your answer by clicking on the link. If your answer is incorrect, watch the video to find your error.

The Ponderal Index measure of leanness states that weight varies directly with the cube of height. If a "normal" person who is 1.2 m tall weighs 21.6 kg, how much will a "normal" person weigh if they are 1.8 m tall?

Section 7.8 Objective 2: Solve Problems Involving Inverse Variation

What is the definition of **Inverse Variation**?

Example 5:
Study the solution for example 5 part a on page 7.8-10 and record the answer below. Complete part b on your own and check your answer by clicking on the link. If your answer is incorrect, watch the video to find your error.

Suppose y varies inversely with x, and $y = 72$ when $x = 50$.

a. Find the equation that relates x and y.

b. Find y when $x = 45$.

Example 6:
Work through example 6 on page 7.8-11 and record the answer below.

For a given mass, the density of an object is inversely proportional to its volume. If 50 cubic centimeters (cm^3), of an object with density of $28g/cm^3$ is compressed to 40 cm^3, what would be its new density?

Example 7:
Work through example 7 on page 7.8-12 and record the answer below. Check your answer by clicking on the link. If your answer is incorrect, watch the video to find your error.

The shutter speed, S, of a camera varies inversely as the square of the aperture setting, f. if the shutter speed is 125 for an aperture of 5.6, what is the shtter speed if the aperture is 1.4?

Section 8.1 Guided Notebook

Section 8.1 Relations and Functions

Read the list of "THINGS TO KNOW" and review any concepts you are unfamiliar with.

Section 8.1 Objective 1: Identify Independent and Dependent Variables

When is a variable **dependent**?

When is a variable **independent**?

Example 1:
Study the solutions for example 1, and record the answers below.

For each of the following equations, identify the dependent variable and the independent variable(s).

a. $y = 3x + 5$ b. $w = ab + 3c^2$ c. $3x^2 + 9y = 12$

Section 8.1 Objective 2: Find the Domain and Range of a Relation

What is a **relation**?

What is the **domain**?

What is the **range**?

Example 2:

Work through example 2 showing all steps below. Check your answers by clicking on the link. If your answers are incorrect, watch the video to find your error.

Find the domain and range of each relation.

a. $\{(-5,7),(3,5),(6,7),(12,-4)\}$

b.

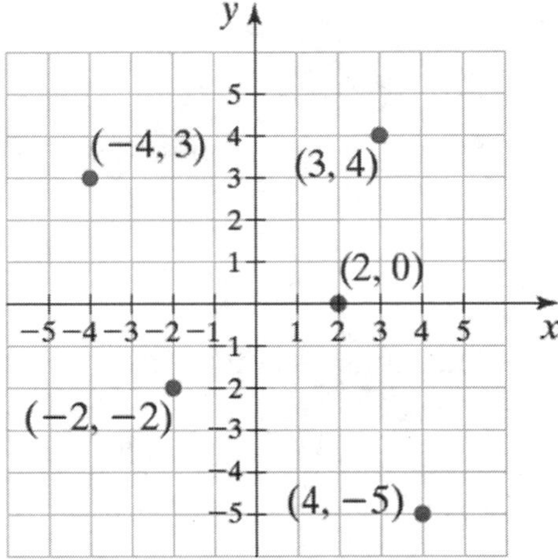

Example 3:

Work through example 3 showing all steps below. Check your answers by clicking on the link. If your answers are incorrect, watch the video to find your error.

a.

b.

c.

What is a **feasible domain**?

Section 8.1 Objective 3: Determine If Relations are Functions

Define a **function**.

How do you determine if a set of ordered pairs is a function?

How do you determine if an equation is a function?

Example 4:
Work through example 4 showing all steps below. Check your answers by clicking on the link. If your answers are incorrect, watch the video to find your error.

Determine if each of the following relations is a function.

a. $\{(-3,6),(2,5),(0,6),(17,-9)\}$

b. $\{(4,5),(7,-3),(4,10),(-6,1)\}$

c. $\{(-2,3),(0,3),(4,3),(6,3),(8,3)\}$

d. $|y-5|=x+3$

e. $y=x^2-3x+2$

f. $4x-8y=24$

Section 8.1 Objective 4: Determine If Graphs Are Functions

What is the **Vertical Line Test** and what is it used for? Watch the animation on 8.1-11.

Example 5:
Work through example 5 showing all steps below. Check your answers by clicking on the link. If your answers are incorrect, watch the animation to find your error.

a. b. c.

d. e. f.

<u>**Section 8.1 Objective 5**: Solve Application Problems Involving Relations and Functions</u>

Example 6:
Study the solutions for example 6 and record the answers below.

The data in the following table represent the average daily hours of sleep and average daily hours of video entertainment for six students at a local college.

Video Entertainment	Sleep	Video Entertainment	Sleep
8	4	5	7
7	5	4	8
2	9	7	6

a. If a researcher believes the number of hours of video entertainment affects the number of hours of sleep, identify the independent variable and the dependent variable.

b. What are the ordered pairs for this data?

c. What are the domain and range?

d. Is this relation a function? Explain.

Example 7:
Study the solution for example 7 part a and record the answer below. Complete parts b - d on your own and check your answers by watching the video.

The percent of households, y, with high-speed internet access in 2007 can be modeled by the equation $y = 0.70x + 20.03$, where x is the annual household income (in \$1000s).

a. Identify the independent and dependent variables.

b. Use the model equation to estimate the percent of households in 2007 with high-speed internet access (to the nearest whole percent) if the annual household income was \$50,000. What point would this correspond to on the graph of the equation?

c. Is the relation a function? Explain.

d. Determine the feasible domain.

Section 8.2 Guided Notebook

Section 8.2 Function Notation and the Algebra of Functions

Read the list of "THINGS TO KNOW" and review any concepts you are unfamiliar with.

Section 8.2 Objective 1: Express Equations of Functions Using Function Notation

What is an example of **function notation** and how is it read?

Read and summarize the CAUTION statement on 8.2-3

What is a benefit of function notation?

Write down the procedure to **Express Equations of Functions Using Function Notation**.

1.

2.

3.

Section 8.2

Read and summarize the CAUTION statement on 8.2-4

Example 1:
Study the solutions for example 1 parts a and b, and record the answers below. Complete part c on your own and check your answer by clicking on the link. If your answer is incorrect watch the video to find your error.

Write each function using function notation. Let x be the independent variable and y be the dependent variable.

a. $y = 2x^2 - 4$ b. $y - \sqrt{x} = 0$ c. $3x + 2y = 6$

__**Section 8.2 Objective 2**: Evaluate Functions__

What does $f(x)$ represent and what is it called?

How do you **evaluate a function**?

Example 2:
Study the solution for example 2 part a, and record the answer below. Complete parts b - d on your own and check your answers by clicking on the link. If your answers are incorrect watch the video to find your error.

If $f(x) = 4x - 5$, $g(t) = 3t^2 - 2t + 1$ and $h(r) = \sqrt{r} - 9$, evaluate each of the following.

a. $f(3)$ b. $g(-1)$ c. $h(16)$ d. $f\left(\frac{1}{2}\right)$

Example 3:
Study the solution for example 3 part a, and record the answer below. Complete parts b – c on your own and check your answers by clicking on the link. If your answers are incorrect watch the video to find your error.

If $P(x) = 4x^3 - 2x^2 + 8x + 7$, evaluate each of the following.

a. $P(4)$ b. $P(-2)$ c. $P\left(-\frac{1}{2}\right)$

Example 4:
Study the solution for example 4 part a, and record the answer below. Complete part b on your own and check your answer by clicking on the link. If your answer is incorrect watch the video to find your error.

If $R(x) = \dfrac{5x^2 - 9}{7x + 3}$, evaluate each of the following.

a. $R(1)$ b. $R(-3)$

<u>Section 8.2 Objective 3:</u> <u>Find the Domain of a Polynomial or Rational Function</u>

What is the **domain** of a polynomial function?

What is the **domain** of a rational function?

What is the procedure for **Finding the Domain of a Rational Function**?

Example 6:
Work through example 6 showing all steps below. Check your answer by clicking on the link. If your answer is incorrect watch the video to find your error.

Find the domain of $g(x) = \dfrac{x^2 + 2x - 15}{x^2 + 5x - 24}$

Example 7:
Work through example 7 part b showing all steps below. Check your answer by clicking on the link. If your answer is incorrect watch the video to find your error.

Find the domain of the rational function.

b. $h(x) = \dfrac{2x - 1}{x^2 + 4}$

<u>**Section 8.2 Objective 4:** Find the Sum, Difference, Product, and Quotient of Functions</u>

Write down the Algebra of Functions:

1.

2.

3.

4.

Example 8:
Study the solution for example 8 part a, and record the answer below. Complete part b on your own and check your answer by clicking on the link. If your answer is incorrect watch the video to find your error.

For $P(x) = x^4 - 9x^2 + 7$ and $Q(x) = 3x^4 - 4x^2 + 2x - 10$, find each of the following.

a. $(P+Q)(x)$

b. $(P-Q)(x)$

Example 9:
Study the solution for example 9 and record the answer below.

For $P(x) = 5x - 2$ and $Q(x) = 4x - 9$, find $(P \cdot Q)(x)$

Example 10:
Work through example 10 showing all steps below. Click on the link to check your answer. If your answer is incorrect, watch the video to find your error.

For $P(x) = 15x^3 + 41x^2 + 4x + 3$ and $Q(x) = 5x + 2$, find $\left(\frac{P}{Q}\right)(x)$. State any values that cannot be included in the domain of $\left(\frac{P}{Q}\right)(x)$ (Note that $Q(x)$ cannot be 0.)

Section 8.3 Guided Notebook

Section 8.3 Function Notation and Applications

Read the list of "THINGS TO KNOW" and review any concepts you are unfamiliar with.

Section 8.3 Objective 1: Graph Simple Functions by Plotting Points

What is the **graph of a function**?

Write down the **Strategy for Graphing Simple Functions by Plotting Points**.

1.

2.

3.

Example 1:
Study the solution for example 1 part a, and record the answer below. Complete parts b - c on your own and check your answers by clicking on the link. If your answers are incorrect watch the animation to find your error.

Graph each function by plotting points.

a. $f(x) = 2x - 1$

b. $g(x) = x^2 + 2x - 3$

c. $h(x) = 2|x| - 1$

Section 8.3 Objective 2: Interpret Graphs of Functions

Example 2:
Study the solutions for example 2.

Example 3:
Work through example 3 showing all steps below. Watch the video to check your answers.

a. Over what time periods was the temperature rising?

b. Over what time periods was the temperature falling?

c. What was the highest temperature for the day? At what time was it reached?

d. What was the lowest temperature for the day shown? At what time was it reached?

e. Over what time period did the temperature decrease most rapidly?

Example 4:
Work through example 4 showing all steps below. Check your answer by clicking on the link. If your answer is incorrect, watch the video to find your error.

A Boeing 757 jet took off and climbed steadily for 20 minutes until it reached an altitude of 18,000 feet. The jet maintained that altitude for 30 minutes. Then it climbed steadily for 10 minutes until it reached an altitude of 26,000 feet. The jet remained at 26,000 feet for 40 minutes. Then it descended steadily for 20 minutes until it reached an altitude of 20,000 feet, where it remained for 30 more minutes. During the final 20 minutes of the flight, the jet descended steadily until it landed at its destination airport. Draw a graph of the 757's altitude as a function of time.

Section 8.3 Objective 3: Solve Application Problems Involving Functions

Example 5:
Study the solutions for example 5 and record the answers below (see video for part f).

A rock is dropped from the top of a cliff. Its height, h, above the ground, in feet at t seconds is given by the function $h(t) = -16t^2 + 900$. Use the model to answer the following questions.

a. Evaluate $h(0)$. What does this value represent?

b. Evaluate $h(2)$. What does this value represent? How far has the rock fallen at this time?

c. Evaluate $h(10)$. Is this possible? Explain.

d. Evaluate $h(7.5)$. Interpret this result.

e. Determine the feasible domain and the range that makes sense (or feasible range) within the context of the problem.

f. Graph the function.

Example 6:
Study the solutions for example 6 and record the answers below (see video for part f).

The average monthly rent, R, for apartments in queens, New York, is modeled by the function $R(a) = 2.2a$, where a is the floor area of the apartment in square feet. Use the model to answer the following questions.

a. What is the average monthly rent for apartments in Queens, New York, with a floor area of 800 square feet?

b. What is the floor area of an apartment if its rent if $1430 per month?

c. Determine the feasible domain and the feasible range of the function.

d. Graph the function.

Section 9.1 Guided Notebook

Section 9.1 Radical Expressions

Read the list of "THINGS TO KNOW" and review any concepts you are unfamiliar with.

Section 9.1 Objective 1: Find Square Roots of Perfect Squares

What is the definition of **Principal and Negative Square Roots**?

Write a radical expression and label all its parts.

When will the square root simplify to a rational number?

Example 1:
Study the solutions for example 1 parts a - c, and record the answers below. Complete parts d - f on your own and check your answers by clicking on the link. If your answers are incorrect watch the video to find your error.

Evaluate.

a. $\sqrt{64}$

b. $-\sqrt{169}$

c. $\sqrt{-100}$

d. $\sqrt{\dfrac{9}{25}}$

e. $\sqrt{0.81}$

f. $\sqrt{0}$

Read and summarize the CAUTION statement on 9.1-5.

Section 9.1 Objective 2: Approximate Square Roots

What happens when the radicand is not a perfect square?

The principal square root of 12 should be between what two numbers? Explain why.

Example 2:
Use your calculator to approximate each square root and round the answer to three decimal places. Check that the answer is reasonable.

a. $\sqrt{5}$ b. $\sqrt{45}$ c. $\sqrt{103}$

Section 9.1 Objective 3: Simplify Radical Expressions of the Form: $\sqrt{a^2}$

Does $\sqrt{a^2}$ always equal a? Explain.

Example 3:
Study the solutions for example 3 parts a - c, and record the answers below. Complete parts d - f on your own and check your answers by clicking on the link. If your answers are incorrect watch the video to find your error.

Simplify.

a. $\sqrt{(-12)^2}$ b. $\sqrt{(2x-5)^2}$ c. $\sqrt{100x^2}$

d. $\sqrt{x^2+12x+36}$ e. $\sqrt{9x^4}$ f. $\sqrt{y^6}$

Section 9.1 Objective 4: Find Cube Roots

What is the **Definition of Cube Roots**?

Can cube roots have negative numbers in the radicand? Why or why not?

Is absolute value used when simplifying cube roots?

Example 4:
Study the solutions for example 4 parts a - c, and record the answers below. Complete parts d - f on your own and check your answers by clicking on the link. If your answers are incorrect watch the video to find your error.

Simplify.

a. $\sqrt[3]{125}$ b. $\sqrt[3]{-1000}$ c. $\sqrt[3]{x^{15}}$

d. $\sqrt[3]{0.064}$ e. $\sqrt[3]{\dfrac{8}{27}}$ f. $\sqrt[3]{-64y^9}$

Read and summarize the CAUTION statement on 9.1-11.

Section 9.1 Objective 5: Find and Approximate *n*th Roots.

What is the **Definition of Principal *n*th Roots**?

What is the **index** of the radical expression and what does it indicate?

Write down the technique for **Simplifying Radical Expressions of the Form:** $\sqrt[n]{a^n}$

Example 5:
Study the solutions for example 5 parts a - c, and record the answers below. Complete parts d - f on your own and check your answers by clicking on the link. If your answers are incorrect watch the video to find your error.

Simplify.

a. $\sqrt[4]{81}$

b. $\sqrt[5]{-32}$

c. $\sqrt[6]{\frac{1}{64}}$

d. $\sqrt[5]{x^{15}}$

e. $\sqrt[6]{(x-7)^6}$

f. $\sqrt[4]{-1}$

Read and summarize the CAUTION statement on 9.1-13.

Example 6:
Use your calculator to approximate each root and round the answer to three decimal places. Check that the answer is reasonable.

a. $\sqrt[3]{6}$

b. $\sqrt[4]{200}$

c. $\sqrt[5]{154}$

Section 9.2 Guided Notebook

Section 9.2 Radical Functions

Read the list of "THINGS TO KNOW" and review any concepts you are unfamiliar with.

Section 9.2 Objective 1: Evaluate Radical Functions

What is a **radical function**?

What is the technique for evaluating radical functions?

Example 1:

Study the solutions for example 1 parts a - c, and record the answers below. Complete parts d - f on your own and check your answers by clicking on the link. If your answers are incorrect watch the video to find your error.

For the radical functions $f(x)=\sqrt{2x-5}$, $g(x)=\sqrt[3]{5x+9}$, and $h(x)=-3\sqrt[4]{x}+2$ evaluate the following.

a. $f(15)$

b. $g(-2)$

c. $h(625)$

d. $g(-\frac{1}{5})$

e. $f(0.5)$

f. $h(1)$

Section 9.2 Objective 2: Find the Domain of a Radical Function.

When can the radicand of a radical expression be negative and when must it be non-negative?

Section 9.2

Write down the **Guideline to Finding the Domain of a Radical Function**.

For an *even* index:

For an *odd* index:

Example 2:
Study the solution for example 2 part a, and record the answer below. Complete parts b - c on your own and check your answers by clicking on the link. If your answers are incorrect watch the video to find your error.

Find the domain for each radical function.

a. $F(x)=\sqrt[4]{12-4x}$ b. $h(x)=\sqrt[5]{3x+5}$ c. $G(x)=\sqrt[6]{5x+7}$

<u>**Section 9.2 Objective 3:** Graph Functions That Contain Square Roots or Cube Roots</u>

What is the **square root function**?

Complete the following chart to find ordered pairs that belong to the function.

x	$y=f(x)=\sqrt{x}$	(x, y)
0		
1		
2		
4		
6		
9		

Why were numbers greater than or equal to zero chosen?

Draw a sketch of the graph showing the ordered pairs of the points plotted.

Example 3:

Study the solution for example 3 part a, and record the answer below. Complete parts b - c on your own and check your answers by clicking on the link. If your answers are incorrect watch the video to find your error.

Graph each function. Compare each graph to that of the square root function.

a. $F(x)=\sqrt{x+1}$ b. $g(x)=\sqrt{x}+1$ c. $f(x)=-\sqrt{x}$

In part a, why were the numbers -1, 0, 3, and 8 chosen for x while in part b the numbers 0, 1, 4, and 9 were chosen for x?

Section 9.2

What is the **cube root function**?

Complete the following chart to obtain ordered pairs that belong to the function.

x	$y = f(x) = \sqrt[3]{x}$	(x, y)
-8		
-1		
0		
1		
8		

Draw a sketch of the graph showing the ordered pairs of the points plotted.

Example 4:
Study the solution for example 4 part a, and record the answer below. Complete parts b - c on your own and check your answers by clicking on the link. If your answers are incorrect watch the video to find your error.

Graph each function. Compare each graph to that of the cube root function.

a. $F(x) = \sqrt[3]{x-2}$ b. $g(x) = \sqrt[3]{x} - 2$ c. $h(x) = -\sqrt[3]{x}$

Section 9.3 Guided Notebook

Section 9.3 Rational Exponents and Simplifying Radical Expressions

Read the list of "THINGS TO KNOW" and review any concepts you are unfamiliar with.

Section 9.3 Objective 1: Use the Definition for Rational Exponents of the Form: $a^{\frac{1}{n}}$

What is the **Definition of a Rational Exponent of the Form** $a^{\frac{1}{n}}$?

Example 1:

Study the solutions for example 1 parts a - c, and record the answers below. Complete parts d - e on your own and check your answers by clicking on the link. If your answers are incorrect watch the video to find your error.

Write each exponential expression as a radical expression. Simplify if possible.

a. $25^{\frac{1}{2}}$

b. $(-64x^3)^{\frac{1}{3}}$

c. $-100^{\frac{1}{2}}$

d. $(-81)^{\frac{1}{2}}$

e. $(7x^3y)^{\frac{1}{5}}$

Example 2:

Study the solution for example 2 part a, and record the answer below. Complete parts b - c on your own and check your answers by clicking on the link. If your answers are incorrect watch the video to find your error.

Write each radical expression as an exponential expression.

a. $\sqrt{5y}$

b. $\sqrt[3]{7x^2y}$

c. $\sqrt[4]{\frac{2m}{3n}}$

Section 9.3 Objective 2: Use the Definition for Rational Exponents of the Form: $a^{\frac{m}{n}}$

What is the **Definition of a Rational Exponent of the Form** $a^{\frac{m}{n}}$ **?**

Example 3:
Study the solutions for example 3 parts a - c, and record the answers below. Complete parts d - e on your own and check your answers by clicking on the link. If your answers are incorrect watch the video to find your error.

Write each exponential expression as a radical expression. Simplify if possible.

a. $16^{\frac{3}{2}}$

b. $\left(\frac{y^3}{1000}\right)^{\frac{2}{3}}$

c. $-81^{\frac{3}{4}}$

d. $(-36)^{\frac{5}{2}}$

e. $(x^2 y)^{\frac{2}{5}}$

Example 4:
Study the solution for example 4 part a, and record the answer below. Complete parts b - c on your own and check your answers by clicking on the link. If your answers are incorrect watch the video to find your error.

Write each radical expression as an exponential expression.

a. $\sqrt[8]{x^5}$

b. $\left(\sqrt[5]{2ab^2}\right)^3$

c. $\sqrt[4]{(10x)^3}$

Example 5:
Study the solutions for example 5 parts a - b. Complete parts c - e on your own and check your answers by clicking on the link. If your answers are incorrect watch the video to find your error.

Write each exponential expression with positive exponents. Simplify if possible.

c. $125^{-\frac{2}{3}}$ d. $\dfrac{1}{8^{-\frac{4}{3}}}$ e. $(-25)^{-\frac{3}{2}}$

<u>Section 9.3 Objective 3:</u> <u>Simplify Exponential Expressions Involving Rational Exponents</u>

Review the **Rules for Exponents** found on page 9.3-12.

Example 6:
Study the solutions for example 6 parts a – c. Complete parts d - f on your own and check your answers by clicking on the link. If your answers are incorrect watch the video to find your error.

Use the rules for exponents to simplify each expression. Assume all variables represent non-negative values.

d. $(32x^{\frac{5}{6}}y^{\frac{10}{9}})^{\frac{3}{5}}$

e. $\left(\dfrac{125x^{\frac{5}{4}}}{y^{\frac{7}{8}}z^{\frac{9}{4}}}\right)^{\frac{4}{3}}$
$y \neq 0, z \neq 0$

f. $\left(4x^{\frac{1}{6}}y^{\frac{3}{4}}\right)^2\left(3x^{\frac{5}{9}}y^{-\frac{3}{2}}\right)$
$y \neq 0$

Section 9.3 Objective 4: Use Rational Exponents to Simplify Radical Expressions

Write down the steps for **Using Rational Exponents to Simplify Radical Expressions:**

1.

2.

3.

Example 7:
Study the solutions for example 7 parts a – c. Complete parts d - f on your own and check your answers by clicking on the link. If your answers are incorrect watch the video to find your error.

Use rational exponents to simplify each radical expression. Assume all variables represent non-negative values.

d. $\sqrt[8]{25x^2y^6}$ e. $\sqrt[4]{49}$ f. $\dfrac{\sqrt[3]{x}}{\sqrt[4]{x}}$ $x \neq 0$

Section 9.3 Objective 5: Simplify Radical Expressions Using the Product Rule

What is the **Product Rule for Radicals**?

Read and summarize the CAUTION statement on 9.3-18.

Write down the steps for **Using the Product Rule to Simplify Radical Expressions of the Form** $\sqrt[n]{a}$

1.

2.

3.

Example 10:

Study the solution for example 10 part a, and record the answer below. Complete parts b - c on your own and check your answers by clicking on the link. If your answers are incorrect watch the video to find your error.

Multiply and simplify. Assume all variables represent non-negative values.

a. $3\sqrt{10}\cdot 7\sqrt{2}$ b. $2\sqrt[3]{4}\cdot 5\sqrt[3]{6}$ c. $\sqrt[4]{18x^3}\cdot\sqrt[4]{45x^2}$

Read and summarize the CAUTION statement on 9.3-24.

<u>**Section 9.3 Objective 6:** Simplify Radical Expressions Using the Quotient Rule</u>

What is the **Quotient Rule for Radicals**?

Section 9.3

Read and summarize the CAUTION statement on 9.3-25.

What are the 3 conditions for a radical expression to be **simplified**?

Condition 1

Condition 2

Condition 3

Example 12:
Study the solutions for example 12 parts a – b, and record the answers below. Complete parts c - d on your own and check your answers by clicking on the link. If your answers are incorrect watch the video to find your error.

Use the quotient rule to simplify. Assume all variables represent positive numbers.

a. $\dfrac{\sqrt{240x^3}}{\sqrt{15x}}$

b. $\dfrac{\sqrt[3]{-500z^2}}{\sqrt[3]{4z^{-1}}}$

c. $\dfrac{\sqrt{150m^9}}{\sqrt{3m}}$

d. $\dfrac{\sqrt{45x^5y^{-3}}}{\sqrt{20xy^{-1}}}$

Section 9.4 Guided Notebook

Section 9.4 Operations with Radicals

Read the list of "THINGS TO KNOW" and review any concepts you are unfamiliar with.

Section 9.4 Objective 1: Add and Subtract Radical Expressions

What is the definition of **Like Radicals**?

Give an example of like radicals and explain why they are like radicals.

Give an example of radicals that are not like radicals and explain why they are not like radicals.

Example 1:
Study the solutions for example 1 and record the answers below.

Add or subtract.

a. $\sqrt{11}+6\sqrt{11}$

b. $7\sqrt{3}-5\sqrt[4]{3}$

c. $\sqrt[3]{\frac{5}{8}}+2\sqrt[3]{5}$

Read and summarize the CAUTION statement on 9.4-8.

Section 9.4

Example 3:
Study the solution for example 3 part a, and record the answer below. Complete part b on your own and check your answer by clicking on the link. If your answer is incorrect watch the video to find your error.

Add or subtract.

a. $\sqrt{54}+6\sqrt{72}-3\sqrt{24}$

b. $\sqrt[3]{24}-\sqrt[3]{192}+4\sqrt[3]{250}$

Example 5:
Work through example 5 showing all steps below. Click on the link to check your answers. If your answers are incorrect, watch the video to find your error.

Add or subtract. Assume variables represent non-negative values.

a. $\dfrac{\sqrt{45}}{6x}-\dfrac{4\sqrt{20}}{5x}$

b. $\dfrac{\sqrt[4]{a^5}}{3}+\dfrac{a\sqrt[4]{a}}{12}$

c. $\dfrac{3x^3\sqrt{24x^3y^3}}{2x\sqrt{3x^2y}}-\dfrac{x^2\sqrt{10xy^4}}{\sqrt{5y^2}}$

Section 9.4 Objective 2: Multiply Radical Expressions

What property is used to multiply rational expressions? Illustrate with example 6 part a.

Example 7:
Study the solution for example 7 part a, and record the answer below. Complete part b on your own and check your answer by clicking on the link.

Multiply. Assume variables represent non-negative values.

a. $\left(7\sqrt{2}-2\sqrt{3}\right)\left(\sqrt{2}-5\right)$

b. $\left(\sqrt{m}-4\right)\left(3\sqrt{m}+7\right)$

Give an example of conjugates that involve radicals.

If the product of conjugates involves square roots, what results? Why?

Section 9.4 Objective 3: Rationalize Denominators of Radical Expressions

What is the procedure to **Rationalize a Denominator with One Term?**

Example 9:
Study the solutions for example 9 and record the answers below.

Rationalize the denominator.

a. $\dfrac{\sqrt{5}}{\sqrt{3}}$

b. $\sqrt{\dfrac{2}{5x}}$

Example 10:
Study the solution for example 10 part a, and record the answer below. Complete part b on your own and check your answer by clicking on the link. If your answer is incorrect watch the video to find your error.

Rationalize the denominator.

a. $\sqrt[3]{\dfrac{11}{25x}}$

b. $\dfrac{\sqrt[4]{7x}}{\sqrt[4]{27y^2}}$

Read and summarize the CAUTION statement on 9.4-19.

Example 11:
Work through example 11 showing all steps below. Click on the link to check your answers. If your answers are incorrect, watch the video to find your error.

Simplify each expression first and then rationalize the denominator.

a. $\sqrt{\dfrac{3x}{50}}$

b. $\dfrac{\sqrt{18x}}{\sqrt{27xy}}$

c. $\sqrt[3]{\dfrac{-4x^5}{16y^5}}$

What is the procedure to **Rationalizing a Denominator with Two Terms**?

Example 12:

Study the solutions for example 12 parts a and b, and record the answers below. Complete part c on your own and check your answer by clicking on the link. If your answer is incorrect watch the video to find your error.

Rationalize the denominator.

a. $\frac{2}{\sqrt{3}+5}$

b. $\frac{7}{3\sqrt{x}-4}$

c. $\frac{\sqrt{y}-3}{\sqrt{y}+2}$

Section 9.5 Guided Notebook

Section 9.5 Radical Equations and Models

Read the list of "THINGS TO KNOW" and review any concepts you are unfamiliar with.

Section 9.5 Objective 1: Solve Equations Involving One Radical Expression

What is the **Definition of a Radical Equation**? Provide two examples.

Read and summarize the CAUTION statement on 9.5-3.

What is the key to solving a radical equation?

How can the *isolated* radical be eliminated from an equation?

What are **extraneous solutions** and when can they occur? Can these solutions be included in the solution set?

Write down the steps for **Solving Equations Involving One Radical Expression**.

1.

2.

3.

4.

Read and summarize the CAUTION statement on 9.5-5

Example 1:
Work through example 1 showing all steps below.

Solve $\sqrt{3x-2}+6=11$.

Read and summarize the CAUTION statement on 9.5-7.

Example 3:
Work through example 3 showing all steps below. Click on the link to check your answer. If your answer is incorrect, watch the video to find your error.

Solve $\sqrt{3x+7}-x=1$

Example 5:
Work through example 5 showing all steps below. Click on the link to check your answer. If your answer is incorrect, watch the video to find your error.

Solve $\left(x^2-9\right)^{\frac{1}{4}}+3=5$

Section 9.5 Objective 2: Solve Equations Involving Two Radical Expressions

Write down the steps for **Solving Equations Involving Two Radical Expressions**.

1.

2.

3.

4.

Example 6:
Work through example 6 showing all steps below. Click on the link to check your answer. If your answer is incorrect, watch the video to find your error.

Solve $\sqrt{x+9}-\sqrt{x}=1$

Example 7:
Work through example 7 showing all steps below. Click on the link to check your answer. If your answer is incorrect, watch the video to find your error.

Solve $\sqrt{2x+3}+\sqrt{x-2}=4$

Section 9.5 Objective 3: Use Radical Equations and Models to Solve Application Problems

Example 9:
Study the solution for example 9 part a, and record the answer below. Complete part b on your own and check your answer by clicking on the link. If your answer is incorrect watch the video to find your error.

Solve each formula for the given variable.

a. Free-falling object: $t = \dfrac{\sqrt{d}}{4}$ for d

b. Radius of a sphere: $r = \sqrt[3]{\dfrac{3V}{4\pi}}$ for V

Example 10:
Study the solution for example 10 part a, and record the answer below. Complete part b on your own and check your answer by clicking on the link. If your answer is incorrect watch the video to find your error.

A SMOG grade for written text is a minimum reading grade level G that a reader must possess in order to fully understand the written text being graded. If w is the number of words that have three or more syllables in a sample of 30 sentences from a given text, the SMOG grade for that text is given by the formula $G = \sqrt{w} + 3$. Use the SMOG grade formula to answer the following questions.

a. If a sample of 30 sentences contains 18 words with three or more syllables, then what is the SMOG grade for the text? Round to a whole number for the grade level.

b. If a text must have a tenth-grade reading level, then how many words with three or more syllables would be needed in the sample of 30 sentences?

Section 9.6 Guided Notebook

Section 9.6 Complex Numbers

Read the list of "THINGS TO KNOW" and review any concepts you are unfamiliar with.

Section 9.6 Objective 1: Simplify Powers of *i*

Why does the equation $x^2 + 1 = 0$ have no real solution? Find the answer by clicking on the link.

Write down the definition of the **Imaginary Unit *i***.

Complete the chart to simplify the powers of *i*:

i	
i^2	
i^3	
i^4	
i^5	
i^6	
i^7	
i^8	

Section 9.6

What is the pattern for the powers of i?

Write down the steps for **Simplifying** i^n **for n>4**.

1.

2.

3.

Example 1:
Study the solutions for example 1 parts a and b, and record the answers below. Complete parts c - e on your own and check your answers by clicking on the link. If your answers are incorrect watch the video to find your error.

Simplify.

a. i^{17} b. i^{60} c. i^{39} d. $-i^{90}$ e. $i^{14} + i^{29}$

What is a **Complex Number**? What is the real part of a complex number? What is the imaginary part of a complex number?

Section 9.6 Objective 2: Add or Subtract Complex Numbers

What is the technique to add or subtract complex numbers?

Example 2:
Study the solutions for example 2 parts a and b. Complete part c on your own and check your answer by clicking on the link. If your answer is incorrect watch the video to find your error.

Perform the indicated operations.

c. $(-3-4i)+(2-i)-(3+7i)$

Section 9.6 Objective 3: Multiply Complex Numbers

What property is used to multiply complex numbers?

Example 4:
Work through example 4 showing all steps below. Click on the link to check your answer. If your answer is incorrect, watch the video to find your error.

Multiply $(4-3i)(7+5i)$

Example 5:
Study the solution for example 5 part a, and record the answer below. Complete part b on your own and check your answer by clicking on the link. If your answer is incorrect watch the video to find your error.

Multiply.

a. $(4+2i)^2$

b. $\left(\sqrt{3}-5i\right)^2$

Section 9.6

Write down the definition of **Complex Conjugates**.

Section 9.6 Objective 4: Divide Complex Numbers

Example 7:
Study the solution for example 7 and record the answer below.

Divide. Write the quotient in stand form. $\dfrac{1-3i}{5-2i}$

Section 9.6 Objective 5: Simplify Radicals with Negative Radicands

What is the rule for the **Square Root of a Negative Number**?

Read and summarize the CAUTION statement on 9.6-20.

Example 10:
Study the solutions for example 10 parts a and b, and record the answers below. Complete parts c - d on your own and check your answers by clicking on the link. If your answers are incorrect watch the video to find your error.

Simplify.

a. $\sqrt{-8}+\sqrt{-18}$

b. $\sqrt{-8}\cdot\sqrt{-18}$

c. $\dfrac{6+\sqrt{(6)^2-4(2)(5)}}{2}$

d. $\dfrac{4-\sqrt{-12}}{4}$

Section 10.1 Guided Notebook

Section 10.1 Solving Quadratic Equations

Read the list of "THINGS TO KNOW" and review any concepts you are unfamiliar with.

Section 10.1 Objective 1: Solve Quadratic Equations Using The Square Root Property

What is a **quadratic equation**?

What is the **Square Root Property**?

Example 1:
Work through example 1 on page 10.1-4 and record the answers below. See how to *check* your answers by clicking on the link.

Use the square root property to solve each quadratic equation. Write each answer in simplest form.

a. $x^2 = 144$ b. $x^2 = 48$

Read and summarize the CAUTION statement on page 10.1-5.

What is the four-step process for **Solving Quadratic Equations Using the Square Root Property**?

1.

2.

3.

4.

Example 2:
Study the solutions for example 2 parts a and b. Complete parts c - d on your own and check your answers by clicking on the link. If your answers are incorrect watch the video to find your error.

Solve

c. $(x-1)^2 = 9$

d. $2(x+1)^2 - 17 = 23$

Section 10.1 Objective 2: Solve Quadratic Equations by Completing the Square

What does it mean to **complete the square**? How do you find the appropriate constant to add?

Example 3:
Study the solutions for example 3 parts a and b. Complete part c on your own and check your answer by clicking on the link. If your answer is incorrect watch the video to find your error.

What number must be added to make the binomial a perfect square trinomial?

c. $x^2 - \frac{3}{2}x$

Write down the steps for **Solving** $ax^2 + bx + c = 0, a \neq 0$ **by Completing the Square**

1.

2.

3.

4.

5.

Example 5:

Study the solution for example 5 and record the answer below.

Solve $2x^2 - 10x - 6 = 0$ by completing the square.

Example 6:

Work through example 6 showing all steps below. Click on the link to check your answer. If your answer is incorrect, watch the video to find your error.

Solve $3x^2 - 18x + 19 = 0$ by completing the square.

Section 10.1 Objective 3: Solve Quadratic Equations Using the Quadratic Formula

Watch the animation on 10.1-15 to see the derivation of the quadratic formula.

What is the **Quadratic Formula**?

Read and summarize the CAUTION statement on 10.1-15.

Example 8:
Work through example 8 showing all steps below. Click on the link to check your answer. If your answer is incorrect, watch the video to find your error.

Solve $3x^2 + 2x - 2 = 0$ using the quadratic formula.

Example 10:
Work through example 10 showing all steps below. Click on the link to check your answer. If your answer is incorrect, watch the video to find your error.

Solve $14x^2 - 5x = 5x^2 + 7x - 4$ using the quadratic formula.

Section 10.1 Objective 4: Use the Discriminant to Determine the Number and Type of Solutions to a Quadratic Equation

Given a quadratic equation $ax^2 + bx + c = 0$ what is the **discriminant**?

Complete the statements below:

If D > 0, then

If D < 0, then

If D = 0, then

Section 10.1 Objective 5: Solve Equations That Are Quadratic in Form

Example 12:
Study the solution for example 12 part a, and record the answer below. Complete parts b - d on your own and check your answers by clicking on the link. If your answers are incorrect watch the video to find your error.

Solve each equation.

a. $2x^4 - 11x^2 + 12 = 0$

b. $\left(\frac{1}{x-2}\right)^2 + \frac{2}{x-2} - 15 = 0$

c. $x^{\frac{2}{3}} - 9x^{\frac{1}{3}} + 8 = 0$

d. $3x^{-2} - 5x^{-1} - 2 = 0$

Section 10.2 Guided Notebook

Section 10.2 Quadratic Functions and Their Graphs

Read the list of "THINGS TO KNOW" and review any concepts you are unfamiliar with.

Section 10.2 Objective 1: Identify the Characteristics of a Quadratic Function from its Graph

What is the definition of a **Quadratic Function**? What is the shape of its graph?

Work through the animation on 10.2-8. Identify each item on the graph and list it.

1. **Vertex** (include *minimum value* and *maximum value* in your description)

2. **Axis of symmetry**

3. **Domain**

4. **Range**

5. **y-intercept**

6. **x-intercept(s)**

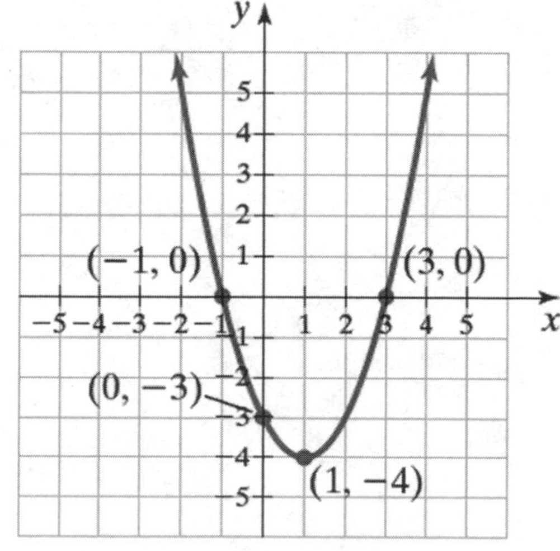

Example 2 graph

Section 10.2 Objective 2: Graph Quadratic Functions by Using Translations

What is a **translation**?

What are **Vertical Shifts of Quadratic Functions**?

What are **Horizontal Shifts of Quadratic Functions**?

Example 5:
Study the solution for example 5 and record the answer below.

Sketch the graph of $g(x)=(x+1)^2+2$. Compare this graph to the graph of $f(x)=x^2$. Show both graphs on the same set of axes.

Section 10.2 Objective 3: Graph Quadratic Functions of the Form: $f(x)=a(x-h)^2+k$

Define the **Standard Form of a Quadratic Function**.

What is another name for standard form?

Section 10.2

Example 7:
Work through example 7 showing all steps below. Click on the link to check your answers. If your answers are incorrect, watch the video to find your error.

Given the quadratic function $f(x) = -(x-2)^2 - 4$, answer each of the following.

a. What are the coordinates of the vertex?

b. Does the graph "open up" or "open down"

c. What is the equation of the axis of symmetry?

d. Find any x-intercepts.

e. Find the y-intercept.

f. Sketch the graph.

g. State the domain and range in interval notation.

Section 10.2 Objective 4: Find the Vertex of a Quadratic Function by Completing the Square

Write down the four steps for **Writing $f(x) = ax^2 + bx + c$ in Standard Form by Completing the Square**.

1.

2.

3.

4.

Section 10.2 Objective 5: Graph Quadratic Functions of the Form: $f(x) = ax^2 + bx + c$ (by Completing the Square)

Example 10:
Work through example 10 showing all steps below. Click on the link to check your answers. If your answers are incorrect, watch the video to find your error.

Rewrite the quadratic function $f(x) = 2x^2 - 4x - 3$ in standard form and then answer the following.

a. What are the coordinates of the vertex?

b. Does the graph “open up” or “open down”

c. What is the equation of the axis of symmetry?

d. Find any x-intercepts.

e. Find the y-intercept.

f. Sketch the graph.

g. State the domain and range in interval notation.

Section 10.2 Objective 6: Find the Vertex of a Quadratic Function by Using the Vertex Formula

Write down the **Formula for the Vertex of a Parabola**.

Example 11:
Study the solution for example 11 part a, and record the answer below. Complete part b on your own and check your answer by clicking on the link. If your answer is incorrect watch the video to find your error.

Use the vertex formula to find the vertex for each quadratic function.

a. $f(x)=3x^2-12x-4$

b. $f(x)=-\frac{1}{2}x^2-10x+5$

<u>**Section 10.2 Objective 7:** Graph Quadratic Functions of the Form</u>: $f(x)=ax^2+bx+c$ (by Using the Vertex Formula).

Example 12:
Work through example 12 showing all steps below. Click on the link to check your answers. If your answers are incorrect, watch the video to find your error.
Given the quadratic function $f(x)=-2x^2-4x+5$, answer the following:

a. What are the coordinates of the vertex?

b. Does the graph "open up" or "open down"

c. What is the equation of the axis of symmetry?

d. Find any x-intercepts.

e. Find the y-intercept.

f. Sketch the graph.

g. State the domain and range in interval notation.

Section 10.3 Guided Notebook

Section 10.3 Applications and Modeling of Quadratic Functions

Read the list of "THINGS TO KNOW" and review any concepts you are unfamiliar with.

Section 10.3 Objective 1: Solve Applications Involving Unknown Numbers

Review the six steps of the **Problem-Solving Strategy for Applications**

Example 2:
Study the solution for example 2 and record the answer below, showing each of the steps of the Problem Solving Strategy.

Three consecutive positive even integers are such that the square of the third is 20 less than the sum of the squares of the first two. Find the positive integers.

Section 10.3 Objective 2: Solve Applications Involving Projectile Motion

Example 3:
Study the solution for example 3 and record the answer below, showing each of the steps of the Problem Solving Strategy.

A toy rocket is launched at an initial velocity of 14.7 m/s from a platform that sits 49 meters above the ground. The height h of the rocket above the ground at any time t seconds after launch is given by the equation $h = -4.9t^2 + 14.7t + 49$. When will the rocket hit the ground?

Section 10.3 Objective 3: Solve Applications Involving Geometric Formulas

Example 4:
Work through example 4 showing each of the steps of the Problem Solving Strategy below. If your answer is incorrect, watch the video to find your error.

The length of a rectangle is 6 inches less than four times the width. Find the dimensions of the rectangle if the area of the rectangle is 54 square inches.

Section 10.3 Objective 4: Solve Applications Involving Distance, Rate, and Time

Example 6:
Work through example 6 showing each of the steps of the Problem Solving Strategy below. Click on the link to check your answer. If your answer is incorrect, watch the video to find your error.

Kevin flew his new Cessna O-2A airplane from Jonesburg to Mountainview, a distance of 2560 miles. The average speed for the return trip was 64 mph faster than the average outbound speed. If the total flying time for the round trip was 18 hours, what was the plane's average speed on the outbound trip from Jonesburg to Mountainview?

Section 10.3 Objective 5: Solve Applications Involving Work

Example 7:
Work through example 7 showing each of the steps of the Problem Solving Strategy below. If your answer is incorrect, watch the video to find your error.

Dawn can finish the monthly sales reports in 2 hours less time than it takes Adam. Working together, they were able to finish the sales reports in 8 hours. How long does it take each person to finish the monthly sales reports alone? (Round to the nearest minute)

Section 10.3 Objective 6: Maximize Quadratic Functions to Solve Application Problems

When does a quadratic function have a **minimum** value and where is it located?

When does a quadratic function have a **maximum** value and where is it located?

Example 8:
Study the solution for example 8 and record the answer below, showing each of the steps of the Problem Solving Strategy.

A toy rocket is launched with an initial velocity of 44.1 meters per second from a platform located 1 meter above the ground. The height h of the object above the ground at any time t seconds after launch is given by the function $h(t) = -4.9t^2 + 44.1t + 1$. How long after launch did it take the rocket to reach its maximum height? What is the rocket's maximum height?

Example 10:
Study the solutions for example 10 parts a and b, and record the answers below. Complete parts c - d on your own and check your answers by clicking on the link. If your answers are incorrect watch the video to find your error.

To sell x waterproof CD alarm clocks, WaterTime, LLC, has determined that the price in dollars must be $p = 250 - 2x$, which is the demand equation. Each clock costs \$2 to produce, with fixed costs of \$4000 per month, producing the cost function $C(x) = 2x + 4000$.

a. Express the revenue R as a function of x.

b. Express the profit P as a function of x.

c. Find the value of x that maximizes profit. What is the maximum profit?

d. What is the price of the alarm clock that will maximize profit?

Section 10.3 Objective 7: Minimize Quadratic Functions to Solve Application Problems

Example 14:
Work through example 14 showing all steps below. Click on the link to check your answers. If your answers are incorrect, watch the video to find your error.

An account rep in one territory oversees $N = 20$ accounts and a second account rep in a nearby territory manages $N = 8$ accounts. The long run average cost function for their industry is $C = N^2 - 70N + 1400$.

a. Determine the long run average cost for $N = 20$ accounts and $N = 8$ accounts.

b. What number of accounts minimizes the long run average cost? What is the minimum long run average cost?

c. Should the two territories be merged into a single territory?

Section 10.4 Guided Notebook

Section 10.4 Circles

Read the list of "THINGS TO KNOW" and review any concepts you are unfamiliar with.

Section 10.4 Objective 1: Find the Distance between Two Points

Watch the video on page 10.4-3 and write down the **Distance Formula**.

Read and summarize the CAUTION statement on 10.4-4.

Example 1:
Work through example 1 showing all steps below.

Find the distance $d(A, B)$ between points (-1, 5) and (4, -5).

Section 10.4 Objective 2: Find the Midpoint of a Line Segment

Write down the **Midpoint of a Line Segment**.

Read and summarize the CAUTION statement on 10.4-6.

Read and summarize the CAUTION statement on 10.4-7.

Example 2:
Work through example 2 showing all steps below.

Find the midpoint of the line segment with endpoints (-3, 2) and (4, 6).

Section 10.4 Objective 3: Write the Standard Form of an Equation of a Circle

Watch the animation on page 10.4-9 and show the steps to develop the **Standard Form of the Equation of a Circle** using the distance formula.

Example 5:
Work through example 5 showing all steps below. Click on the link to check your answer. If your answer is incorrect, watch the video to find your error.

Write the standard form of the equation of the circle with center (0, -4) and radius $r = \sqrt{5}$.

<u>**Section 10.4 Objective 4:** Sketch the Graph of a Circle Given in Standard Form</u>

Example 8:
Work through example 8 showing all steps below. Click on the link to check your answer. If your answer is incorrect, watch the video to find your error.

Find the center and the radius, and sketch the graph of the circle $(x-1)^2 + (y+2)^2 = 9$. Also find any intercepts.

<u>**Section 10.4 Objective 5:** Write the General Form of a Circle in Standard Form and Sketch Its Graph</u>

What is the **General Form of the Equation of a Circle**?

What procedure is used to change from the general form of a circle to the standard form of a circle?

Example 10:
Work through example 10 showing all steps below. If your answer is incorrect watch the video to find your error.

Write the equation $x^2 + y^2 - 8x + 6y + 16 = 0$ in standard form; find the center, radius, and intercepts, and sketch the graph.

Example 11:
Watch the animation for example 11 and answer the question below.

Write the equation $4x^2 + 4y^2 + 4x - 8y + 1 = 0$ in standard form; find the center, radius, and intercepts, and sketch the graph.

Section 10.5 Guided Notebook

Section 10.5 Polynomial and Rational Inequalities

Read the list of "THINGS TO KNOW" and review any concepts you are unfamiliar with.

Section 10.5 Objective 1: Solve Polynomial Inequalities

Define a **Polynomial Inequality**.

What play an important part when solving polynomial inequalities?

Example 1:
Watch the video for example 1 and answer the question below. Record the seven steps for **Solving Polynomial Inequalities** as you work the problem.

Solve $x^3 - 3x^2 + 2x \geq 0$

1.

2.

3.

4.

Example 1: continued

5.

6.

7.

Example 2:
Work through example 2 showing all steps below.

$x^2 + 5x < 3 - x^2$

Section 10.5 Objective 2: Solve Rational Inequalities

Define a **Rational Inequality**.

Section 10.5

Write down the steps for **Solving Rational Inequalities.**

1.

2.

3.

4.

5.

6.

7.

Example 3:

Work through example 3 showing all steps below.

Solve $\frac{x-4}{x+1} \geq 0$

Read and summarize the CAUTION statement on 10.5-18.

Example 4:
Work through example 4 showing all steps below.

Solve $x > \dfrac{3}{x-2}$

Example 5:
Work through example 5 showing all steps below. Click on the link to check your answer. If your answer is incorrect, watch the video to find your error.

Solve $\dfrac{x+1}{x-2} > \dfrac{7x+1}{x^2+x-6}$

Section 11.1 Guided Notebook

Section 11.1 Transformations of Functions

Read the list of "THINGS TO KNOW" and review any concepts you are unfamiliar with.

Section 11.1 Objective 1: Use Vertical Shifts to Graph Functions

Click on the **Basic Functions** to review the properties of each graph.

Example 1:
Work through example 1 showing all steps below.

Sketch the graphs of $f(x)=|x|$ and $g(x)=|x|+2$.

Write down the information for **Vertical Shifts of Functions**.

Section 11.1 Objective 2: Use Horizontal Shifts to Graph Functions

Sketch the graphs of $f(x) = x^2$ and $g(x) = (x+2)^2$.

Write down the information for **Horizontal Shifts of Functions**

Example 2:
Watch the animation for example 2 and show all work below.

Use the graph of $y = x^3$ to sketch the graph of $g(x) = (x-1)^3 + 2$.

Section 11.1 Objective 3: Use Reflections to Graph Functions

Write down the information about **Reflections of Functions about the x-Axis**. Watch the animation.

Write down the information about **Reflections of Functions about the y-Axis.** Watch the animation.

Section 11.1 Objective 4: Use Vertical Stretches and Compressions to Graph Functions

Write down the information about **Vertical Stretches and Compressions of Functions**. Watch the animations.

Section 11.1 Objective 5: Use Horizontal Stretches and Compressions to Graph Functions

Write down the information about **Horizontal Stretches and Compressions of Functions**. Watch the animations.

Section 11.1 Objective 6: Use Combinations of Transformations to Graph Functions

Write down the "order of operations" for sketching a function that involves multiple transformations.

1.

2.

3.

4.

5.

6.

Example 6:
Watch the animation for example 6 and show all work below.

Use transformations to sketch the graph of $f(x) = -2(x+3)^2 - 1$.

Example 7:
Work through example 7 showing all work below. If your graphs are incorrect watch the video to find your error.

Draw the graph on page 11.1-25 and *label* the key ordered pairs.

Use the above graph of $y = f(x)$ to sketch each of the following functions.

a. $y = -f(2x)$

b. $y = 2f(x-3) - 1$

c. $y = -\frac{1}{2}f(2-x)+3$

Section 11.2 Guided Notebook

Section 11.2 Composite and Inverse Functions

Read the list of "THINGS TO KNOW" and review any concepts you are unfamiliar with.

Section 11.2 Objective 1: Form and Evaluate Composite Functions

Begin learning objective 1 by watching the video on 11.2-3.

Read and summarize the CAUTION statement on 11.2-4.

Example 1:

Work through example 1 showing all steps below. If your answers are incorrect, watch the video to find your error.

Let $f(x) = 4x + 1$, $g(x) = \dfrac{x}{x-2}$ and $h(x) = \sqrt{x+3}$.

a. Find the function $f \circ g$

b. Find the function $g \circ h$

c. Find the function $h \circ f \circ g$

d. Evaluate $(f \circ g)(4)$ or state that it is undefined.

e. Evaluate $(g \circ h)(1)$ or state that it is undefined.

f. Evaluate $(h \circ f \circ g)(6)$ or state that it is undefined.

Section 11.2

Example 2:
Work through example 2 showing all steps below. If your answers are incorrect, watch the video to find your error.

Use the graph to evaluate each expression.

a. $(f \circ g)(4)$ b. $(g \circ f)(-3)$ c. $(f \circ f)(-1)$

d. $(g \circ g)(4)$ e. $(f \circ g \circ f)(1)$

Section 11.2 Objective 2: Determine the Domain of Composite Functions

Example 3:
Work through example 3 showing all steps below. If your answers are incorrect, watch the video to find your error.

Let $f(x) = \dfrac{-10}{x-4}$ and $g(x) = \sqrt{5-x}$.

a. Find the domain of $f \circ g$ b. Find the domain of $g \circ f$

Section 11.2 Objective 3: Determine If a Function is One-to-One Using the Horizontal Line Test

Begin learning objective 3 by watching the video on 11.2-11.

What is the definition of a **One-to-One Function**?

Example 4:
Work through example 4 parts a - c showing all steps below. If your answers are incorrect, watch the animation to find your error.

Determine whether each function is one-to-one.

a. Use the given graph on 11.2-15

b. Use the given graph on 11.2-15

c. $f(x) = x^2 + 1, x \leq 0$

Section 11.2 Objective 4: Verify Inverse Functions

What is the definition of **Inverse Function**?

The ______________ of f is exactly the same as the ______________ of f^{-1}, and the ______________ of f is exactly the same as the ______________ of f^{-1}.

If the point (a, b) is on the graph of f then the point _________ is on the graph of f^{-1}.

Read and summarize the CAUTION statement on 11.2-18.

Example 5:
Work through example 5 showing all steps below. If your answer is incorrect, watch the video to find your error.

Show that $f(x) = \dfrac{x}{2x+3}$ and $g(x) = \dfrac{3x}{1-2x}$ are inverse functions using the composition cancellation equations.

Section 11.2 Objective 5: Sketch the Graphs of Inverse Functions

Given the graph of a one-to-one function, how can we obtain the graph of its inverse?

Example 6:
Work through example 6 showing all steps below. If your answers are incorrect, watch the animation to find your error.

Sketch the graph of $f(x) = x^2 + 1$, $x \le 0$ and its inverse. Also state the domain and range of f and f^{-1}.

Section 11.2 Objective 6: Find the Inverse of a One-to-One Function

Write down the four steps to find the inverse of a one-to-one function.

1.

2.

3.

4.

Example 7:
Study the animation for example 7.

Find the inverse of the function $f(x) = \dfrac{2x}{1-5x}$, and state the domain and range of f and f^{-1}.

Section 11.3 Guided Notebook

Section 11.3 Exponential Functions

Read the list of "THINGS TO KNOW" and review any concepts you are unfamiliar with.

Section 11.3 Objective 1: Use the Characteristics of Exponential Functions

Write down the definition of an **Exponential Function**.

Complete the following table.

x	$y=2^x$	$y=3^x$	$y=\left(\frac{1}{2}\right)^x$	$y=\left(\frac{1}{3}\right)^x$
-2				
-1				
0				
1				
2				

What point do all four graphs have in common? Why is this?

Example 1:
Work through example 1 showing all steps below. If your answer is incorrect, watch the video to find your error.

Sketch the graph of $f(x)=\left(\frac{2}{3}\right)^x$

Example 2:
Study the solution for example 2, and record the answer below.

Find the exponential function $f(x)=b^x$ whose graph is given on 11.3-9.

<u>Section 11.3 Objective 2:</u> <u>Sketch the Graphs of Exponential Functions Using Transformations</u>

Example 3:
Work through example 3 showing all steps below. If your answer is incorrect, watch the video to find your error.

Use transformations to sketch the graph of $f(x)=-2^{x+1}+3$.

How do you find the value of the y-intercept?

How do you find the value of the x-intercept?

<u>Section 11.3 Objective 3:</u> <u>Solve Exponential Equations by Relating the Bases</u>

If the bases of an exponential equation are the same what can be said about the exponents?

Write down the **Method of Relating the Bases**.

Example 4:
Work through example 4 showing all steps below. If your answers are incorrect, watch the animation to find your error.

Solve the following equations.

a. $8 = \frac{1}{16^x}$

b. $\frac{1}{27^x} = \left(\sqrt[4]{3}\right)^{x-2}$

<u>**Section 11.3 Objective 4:** Solve Applications of Exponential Functions</u>

Example 5:
Work through example 5 showing all steps below.

Most golfers find that their golf skills improve dramatically at first and level off rather quickly. For example, suppose that the distance (in yards) that a typical beginning golfer can hit a 3-wood after t weeks of practice on the driving range is given by the exponential function $d(t) = 225 - 100(2.7)^{-0.7t}$. This function has been developed after many years of gathering data on beginning golfers.

How far can a typical beginning golfer initially hit a 3-wood? How far can a typical beginning golfer hit a 3-wood after 1 week of practice on the driving range? After 5 weeks? After 9 weeks? Round to the nearest hundredth yard.

Section 11.3

Write down the **Periodic Compound Interest Formula**. Be sure to state what each variable represents.

Example 6:
Work through example 6 showing all steps below.

Which investment will yield the most money after 25 years?

Investment A: $12,000 invested at 3% compounded monthly
Investment B: $10,000 invested at 3.9% compounded quarterly

Write down the **Present Value Formula**. Be sure to state what each variable represents.

Section 11.4 Guided Notebook

Section 11.4 The Natural Exponential Function

Read the list of "THINGS TO KNOW" and review any concepts you are unfamiliar with.

Section 11.4 Objective 1: Use the Characteristics of the Natural Exponential Function

Begin learning objective 1 by watching the video on 11.4-3.

What is the definition of **Natural Base**?

Example 1:
Work through example 1 with your calculator.

Evaluate each expression correctly to six decimal places.

a. e^2 b. $e^{-0.534}$ c. $1000e^{0.013}$

Section 11.4 Objective 2: Sketch the Graphs of Natural Exponential Functions Using Transformations

Example 2:
Work through example 2 showing all steps below. If your answers are incorrect, watch the video to find your error.

Use transformations to sketch the graph of $f(x) = -e^x + 2$. Determine the domain, range, and y-intercept and find the equation of any asymptotes.

Section 11.4 Objective 3: Solve Natural Exponential Equations by Relating the Bases

Example 3:
Work through example 3 showing all steps below. If your answers are incorrect, watch the video to find your error.

Use the method of relating the bases to solve each exponential equation.

a. $e^{3x-1} = \dfrac{1}{\sqrt{e}}$

b. $\dfrac{e^{x^2}}{e^{10}} = \left(e^{x}\right)^3$

Section 11.4 Objective 4: Solve Applications of the Natural Exponential Function

Write down the **Continuous Compound Interest Formula**. Be sure to state what each variable represents.

Example 4:
Work through example 4 showing all steps below.

How much money would be in an account after 5 years if an original investment of \$6000 was compounded continuously at 4.5%? Compare this amount to the same investment that was compounded daily. Round to the nearest cent.

Write down the **Present Value Formula**. Be sure to state what each variable represents.

Example 5:
Work through example 5 showing all steps below.

Find the present value of \$18,000 if interest is paid at a rate of 8% compounded continuously for 20 years. Round to the nearest cent.

Write down the model for **Exponential Growth.** Be sure to state what each variable represents.

Example 6:
Work through example 6 showing all steps below.

The population of a small town follows the exponential growth model $P(t) = 900e^{0.015t}$, where t is the number of years after 1900.

Answer the following questions, rounding each answer to the nearest whole number:

a. What was the population of this town in 1900?

b. What was the population of this town in 1950?

c. Use this model to predict the population of this town in 2012.

Example 7:
Work through example 7 showing all steps below.

Twenty years ago, the State of Idaho Fish and Game Department introduced a new breed of wolf into a certain Idaho forest. The current wolf population in this forest is now estimated at 825, with a relative growth rate of 12%.

Answer the following questions, rounding each answer to the nearest whole number:

a. How many wolves did the Idaho Fish and Game Department initially introduce into this forest?

b. How many wolves can be expected after another 20 years?

Section 11.5 Guided Notebook

Section 11.5 Logarithmic Functions

Read the list of "THINGS TO KNOW" and review any concepts you are unfamiliar with.

Section 11.5 Objective 1: Use the Definition of a Logarithmic Function

Begin learning objective 1 by watching the video on 11.5-3.

Write down the definition of a **Logarithmic Function**.

Write down steps to find the inverse of $f(x) = b^x$.

1.

2.

3.

4.

Example 1:
Work through example 1 showing all steps below. If your answers are incorrect, watch the video to find your error.

Write each exponential equation as an equation involving a logarithm.

a. $2^3 = 8$ b. $5^{-2} = \frac{1}{25}$ c. $1.1^M = z$

Example 2:
Work through example 2 showing all steps below. If your answers are incorrect, watch the video to find your error.

Write each logarithmic equation as an equation involving an exponent.

a. $\log_3 81 = 4$

b. $\log_4 16 = y$

c. $\log_{3/5} x = 2$

Section 11.5 Objective 2: Evaluate Logarithmic Expressions

What are two ways logarithms may be evaluated?

Example 3:
Work through example 3 showing all steps below. If your answers are incorrect, watch the video to find your error.

Evaluate each logarithm:

a. $\log_5 25$

b. $\log_3 \frac{1}{27}$

c. $\log_{\sqrt{2}} \frac{1}{4}$

Section 11.5 Objective 3: Use the Properties of Logarithms

Write down the **General Properties of Logarithms**.

Write down the **Cancellation Properties of Exponentials and Logarithms**.

Example 4:
Study the solutions for example 4, and record the answers below.

Use the properties of logarithms to evaluate each expression.

a. $\log_3 3^4$ b. $\log_{12} 12$ c. $7^{\log_7 13}$ d. $\log_8 1$

Section 11.5 Objective 4: Use the Common and Natural Logarithms

What is a **common logarithm**?

What is a **natural logarithm**?

Example 5:
Work through example 5 showing all steps below. If your answers are incorrect, watch the video to find your error.

Write each exponential equation as an equation involving a common logarithm or natural logarithm.

a. $e^0 = 1$

b. $10^{-2} = \frac{1}{100}$

c. $e^K = w$

Example 6:
Work through example 6 showing all steps below. If your answers are incorrect, watch the video to find your error.

Write each logarithmic equation as an equation involving an exponent.

a. $\log 10 = 1$

b. $\ln 20 = Z$

c. $\log(x-1) = T$

Example 7:
Work through example 7 showing all steps below. If your answers are incorrect, watch the video to find your error.

Evaluate each expression without the use of a calculator.

a. $\log 100$

b. $\ln\sqrt{e}$

c. $e^{\ln 51}$

d. $\log 1$

Section 11.5 Objective 5: Use the Characteristics of Logarithmic Functions

Write down the three steps to sketch the graph of a logarithmic function.

1.

2.

3.

Example 8:
Study the video for example 8.

Sketch the graph of $f(x) = \log_3 x$.

<u>**Section 11.5 Objective 6:** Sketch the Graphs of Logarithmic Functions Using Transformations</u>

Example 9:
Work through example 9 showing all steps below. If your answer is incorrect, watch the video to find your error.

Sketch the graph of $f(x) = -\ln(x+2) - 1$

<u>**Section 11.5 Objective 7:** Find the Domain of Logarithmic Functions</u>

Find the domain of $f(x) = -\ln(x+2) - 1$

Example 10:
Work through example 10 showing all steps below. If your answer is incorrect, watch the video to find your error.

Find the domain of $f(x) = \log_5\left(\frac{2x-1}{x+3}\right)$

Section 11.6 Guided Notebook

Section 11.6 Properties of Logarithms

Read the list of "THINGS TO KNOW" and review any concepts you are unfamiliar with.

Section 11.6 Objective 1: Use the Product Rule, Quotient Rule, and Power Rule for Logarithms

Write down the **Properties of Logarithms**. Watch any *one* of the videos for a proof.

1.

2.

3.

Example 1:
Study the solutions for example 1, and record the answers below.

Use the product rule for logarithms to expand each expression. Assume x > 0.

a. $\ln(5x)$ b. $\log_2(8x)$

Read and summarize the CAUTION statement on 11.6-4.

Example 2:
Study the solutions for example 2, and record the answers below.

Use the quotient rule for logarithms to expand each expression. Assume $x > 0$.

a. $\log_5\left(\frac{12}{x}\right)$

b. $\ln\left(\frac{x}{e^5}\right)$

Read and summarize the CAUTION statement on 11.6-5

Example 3:
Study the solutions for example 3, and record the answers below.

Use the power rule for logarithms to expand each expression. Assume $x > 0$.

a. $\log 6^3$

b. $\log_{\frac{1}{2}} \sqrt[4]{x}$

Read and summarize the CAUTION statement on 11.6-6.

Section 11.6

Section 11.6 Objective 2: Expand and Condense Logarithmic Expressions

Example 4:
Work through example 4 part b showing all steps below. If your answer is incorrect, watch the video to find your error.

Use properties of logarithms to expand the logarithmic expression as much as possible.

b. $\ln\left(\frac{(x^2-4)}{9e^{x^3}}\right)$

Example 5:
Work through example 5 part b showing all steps below. If your answer is incorrect, watch the video to find your error.

Use the properties of logarithms to rewrite the expression as a single logarithm.

b. $\frac{1}{3}(\log_3 x - 2\log_3 y) + \log_3 10$

Section 11.6 Objective 3: Solve Logarithmic Equations Using the Logarithm Property of Equality

Write down the **Logarithm Property of Equality**.

Example 6:
Study the solutions for example 6, and record the answers below.

Solve the following equations.

a. $\log_7(x-1) = \log_7 12$

b. $2\ln x = \ln 16$

Read and summarize the CAUTION statement on 11.6-13.

Section 11.6 Objective 4: Use the Change of Base Formula

Write down the **Change of Base Formula**. Watch the video for the proof of the change of base formula.

Example 7:
Study the solutions for example 7, and record the answers below.

Approximate the following expressions. Use a common log for part a and use a natural log for part b. Round each to four decimal places.

a. $\log_9 200$

b. $\log_{\sqrt{3}} \pi$

Example 8:
Study the solution for example 8, and record the answer below.

Use the change of base formula and the properties of logarithms to rewrite as a single logarithm involving base 2.

$\log_4 x + 3\log_2 y$

Redo example 8. This time use the change of base formula and the properties of logarithms to rewrite as a single logarithm involving **base 4**. Watch the video if your answer is incorrect.

Example 9:
Work through example 9 showing all steps below. If your answer is incorrect, watch the video to find your error.

Use the change of base formula and the properties of logarithms to solve the equation.

$2\log_3 x = \log_9 16$

Section 11.7 Guided Notebook

Section 11.7 Exponential and Logarithmic Equations

Read the list of "THINGS TO KNOW" and review any concepts you are unfamiliar with.

Section 11.7 Objective 1: Solve Exponential Equations

Recall the **logarithm property of equality**:

Recall the **power rule for logarithms**:

Example 1:
Study the solution for example 1, showing all steps below. Watch the video for further explanation, if needed.

Solve $2^{x+1} = 3$

Write down the procedure for **Solving Exponential Equations**.

Example 2:
Work through example 2 showing all steps below. Click on the link to check your answers. If your answers are incorrect, watch the video to find your error.

Solve each equation. For part b, round to four decimal places.

a. $3^{x-1} = \left(\frac{1}{27}\right)^{2x+1}$

b. $7^{x+3} = 4^{2-x}$

Example 3:
Study the solution for example 3, showing all steps below. Watch the video for further explanation, if needed.

Solve each equation. Round to four decimal places.

a. $25e^{x-5} = 17$

b. $e^{2x-1} \cdot e^{x+4} = 11$

Section 11.7 Objective 2: Solve Logarithmic Equations

Recall the **Properties of Logarithms**:

1.

2.

3.

Example 4:
Study the solution for example 4, showing all steps below. Watch the video for further explanation, if needed.

Solve $2\log_5(x-1)=\log_5 64$

Why must the solution $x=-7$ be discarded?

Read and summarize the CAUTION statement on 11.7-13.

Section 11.7

Write down the steps for **Solving Logarithmic Equations**.

1.

2.

3.

4.

5.

Example 5:
Study the solution for example 5, showing all steps below. Watch the video for further explanation, if needed.

Solve $\log_4(2x-1)=2$

Example 6:
Study the solution for example 6, showing all steps below. Watch the video for further explanation, if needed.

Solve $\log_2(x+10)+\log_2(x+6)=5$

Since both answers are negative, should they both be discarded? Explain.

Example 7:
Study the solution for example 7, showing all steps below.

Solve $\ln(x-4)-\ln(x-5)=2$ Round to four decimal places.

Section 11.8 Guided Notebook

Section 11.8 Applications of Exponential and Logarithmic Functions

Read the list of "THINGS TO KNOW" and review any concepts you are unfamiliar with.

Section 11.8 Objective 1: Solve Compound Interest Applications

Recall the **Periodic Compound Interest Formula** and the **Continuous Compound Interest Formula**. Be sure to state what each variable represents.

Example 1:
Study the solution for example 1, showing all steps below. Watch the video for further explanation, if needed.

How long will it take (in years and months) for an investment to double if it earns 7.5% compounded monthly?

Example 2:
Study the solution for example 2, showing all steps below. Watch the video for further explanation, if needed.

Suppose an investment of $5000 compounded continuously grew to an amount of $5130.50 in 6 months. Find the interest rate, and then determine how long it will take for the investment to grow to $6000. Round the interest rate to the nearest hundredth of a percent and the time to the nearest hundredth of a year.

Section 11.8 Objective 2: Solve Exponential Growth and Decay Applications

Write down the **Exponential Growth** model. Be sure to state what each variable represents.

Section 11.8

Example 3:

Study the solution for example 3, showing all steps below. Watch the video for further explanation, if needed.

The population of a small town grows at a rate proportional to its current size. In 1900, the population was 900. In 1920, the population had grown to 1600. What was the population of this town in 1950? Round to the nearest whole number.

Write down the **Exponential Decay** model. Be sure to state what each variable represents.

Study the animation for Half-Life on page 11.8-13.

Example 4:
Study the solution for example 4, showing all steps below. Watch the video for further explanation, if needed.

Suppose that a meteorite is found containing 4% of its original Krypton-99. If the half-life of Krypton-00 is 80 years, how old is the meteorite? Round to the nearest year.

Section 11.8 Objective 3: Solve Logistic Growth Applications

Write down the **Logistic Growth** model. Be sure to state what each variable represents.

Example 5:
Study the solution for example 5, showing all steps below. Watch the video for further explanation, if needed.

Ten goldfish were introduced into a small pond. Because of limited food, space, and oxygen, the carrying capacity of the pond is 400 goldfish. The goldfish population at any time t, in days, is modeled by the logistic growth function $F(t)=\dfrac{C}{1+Be^{kt}}$. If 30 goldfish are in the pond after 20 days,

a. Find B.

b. Find k.

c. When will the pond contain 250 goldfish? Round to the nearest whole number.

Section 11.8 Objective 4: Use Newton's Law of Cooling

Write down the model for **Newton's Law of Cooling**. Be sure to state what each variable represents.

Example 6:

Study the solution for example 6, showing all steps below. Watch the video for further explanation, if needed.

Suppose that the temperature of a cup of hot tea obeys Newton's law of cooling. If the tea has a temperature of 200° F when it is initially poured and 1 minute later has cooled to 189°F in a room that maintains a constant temperature of 69° F, determine when the tea reaches a temperature of 146°F. Round to the nearest minute.

Section 12.1 Guided Notebook

Introduction to Conics Sections

View the animations on each of the four conic sections, making note of how each is formed.

Section 12.1 The Parabola

Read the list of "THINGS TO KNOW" and review any concepts you are unfamiliar with.

Section 12.1 Objective 1: Work with the Equation of a Parabola with a Vertical Axis of Symmetry

View the animation on page 12.1-3 and describe the **Characteristics of a Parabola**. Take detailed notes below.

In section 8.2 you studied the algebraic definition of a parabola. What is the **Geometric Definition of the Parabola**?

Section 12.1

Write down the **Equation of a Parabola in Standard Form with a Vertical Axis of Symmetry**.

Draw two sketches of a vertical parabola.

What is the vertex?

What is the distance from the vertex to the focus?

What is the distance from the vertex to the directrix?

What is the focus?

What is the equation of the directrix?

Example 1:
Study the solution for example 1, showing all work below.

Find the vertex, focus and directrix of the parabola $x^2 = 8y$ and sketch its graph.

Example 2:
Study the solution for example 2, showing all steps below. Watch the video for further explanation, if needed.

Find the vertex, focus, and directrix of the parabola $-(x+1)^2 = 4(y-3)$ and sketch its graph.

Section 12.1 Objective 2: Work with the Equation of a Parabola with a Horizontal Axis of Symmetry

Write down the **Equation of a Parabola in Standard Form with a Horizontal Axis of Symmetry**.

Draw two sketches of a horizontal parabola.

What is the vertex?

What is the distance from the vertex to the focus?

What is the distance from the vertex to the directrix?

What is the focus?

What is the equation of the directrix?

Example 3:
Study the solution for example 3, showing all steps below. Watch the video for further explanation, if needed.

Find the vertex, focus, and directrix of the parabola $(y-3)^2 = 8(x+2)$

<u>**Section 12.1 Objective 3:** Find the Equation of a Parabola Given Information about the Graph</u>

Example 4:
Study the solution for example 4, showing all steps below. Watch the video for further explanation, if needed.

Find the standard form of the equation of the parabola with focus $\left(-3, \frac{5}{2}\right)$ and directrix $y = \frac{11}{2}$

Example 5:
Study the solution for example 5, showing all steps below. Watch the video for further explanation, if needed.

Find the standard form of the equation of the parabola with focus $(4,-2)$ and vertex $\left(\frac{13}{2}, -2\right)$

Section 12.1 Objective 4: Complete the Square to Find the Equation of a Parabola in Standard Form

Example 6:
Study the solution for example 6, showing all steps below. Watch the video for further explanation, if needed.

Find the vertex, focus, and directrix and sketch the graph of the parabola $x^2 - 8x + 12y = -52$

Section 12.1 Objective 5: Solving Applications Involving Parabolas

Example 7:
Study the solution for example 7, showing all steps below.

Parabolic microphones can be seen on the sidelines of professional sporting events so that television networks can capture audio sounds from the players on the field. If the surface of a parabolic microphone is 27 centimeters deep and has a diameter of 72 centimeters at the top, where should the microphone be placed relative to the vertex of the parabola?

Section 12.2 Guided Notebook

Section 12.2 The Ellipse

Read the list of "THINGS TO KNOW" and review any concepts you are unfamiliar with.

Section 12.2 Objective 1: Sketch the Graph of an Ellipse

What is the **Geometric Definition of the Ellipse?**

Watch the video on 12.2-5 describing horizontal and vertical ellipses. Draw a sketch and label the key features of each below.

Write down the **Equation of an Ellipse in Standard Form with Center (h,k) and a Horizontal Major Axis**.

What is the relationship between a and b?

What are the ordered pairs of the foci?

What are the order pairs of the vertices?

What are the endpoints of the major axis?

What is the relationship between a, b, and c?

Write down the **Equation of an Ellipse in Standard Form with Center (h,k) and a Vertical Major Axis**.

What is the relationship between a and b?

What are the ordered pairs of the foci?

What are the order pairs of the vertices?

What are the endpoints of the major axis?

What is the relationship between a, b, and c?

Example 1:

Study the solution for example 1, showing all steps below. Watch the video for further explanation, if needed.

Sketch the graph of the following ellipse and label the center, foci, and vertices.

$$\frac{x^2}{25}+\frac{y^2}{4}=1$$

Example 2:
Study the solution for example 2, showing all steps below. Watch the video for further explanation, if needed.

Sketch the graph of the following ellipse and label the center, foci, and vertices.

$$\frac{(x+2)^2}{20}+\frac{(y-3)^2}{36}=1$$

<u>**Section 12.2 Objective 2:** Find the Equation of an Ellipse Given Information about the Graph</u>

Example 3:
Study the solution for example 3, showing all steps below. Watch the video for further explanation, if needed.

Find the standard form of the equation of the ellipse with foci at (-6,1) and (-2,1) such that the length of the major axis is eight units.

Example 4:
Study the solution for example 4, showing all steps below. Watch the video for further explanation, if needed.

Determine the equation of the ellipse with foci located at (0,6) and (0,-6) that passes through the point (-5,6)

<u>**Section 12.2 Objective 3:** Complete the Square to Find the Equation of an Ellipse in Standard Form</u>

Example 5:
Study the solution for example 5, showing all steps below. Watch the video for further explanation, if needed.

Find the center and foci and sketch the ellipse

$36x^2 + 20y^2 + 144x - 120y - 396 = 0$

Section 12.2 Objective 4: Applications Involving Ellipses

Example 6:
Study the solution for example 6, showing all steps below.

A patient is placed in an elliptical tank that is 280 centimeters long and 250 centimeters wide to undergo sound wave lithotripsy treatment for kidney stones. Determine where the sound emitter and the stone should be positioned relative to the center of the ellipse.

Section 12.3 Guided Notebook

Section 12.3 The Hyperbola

Read the list of "THINGS TO KNOW" and review any concepts you are unfamiliar with.

Section 12.3 Objective 1: Sketch the Graph of a Hyperbola

What is the **Geometric Definition of the Hyperbola**?

Draw a sketch of two different hyperbolas below.

Identify each of the following in your sketches above.

Center	Foci	Transverse axis
Vertices	Asymptotes	Conjugate Axis
Reference rectangle		

Section 12.3

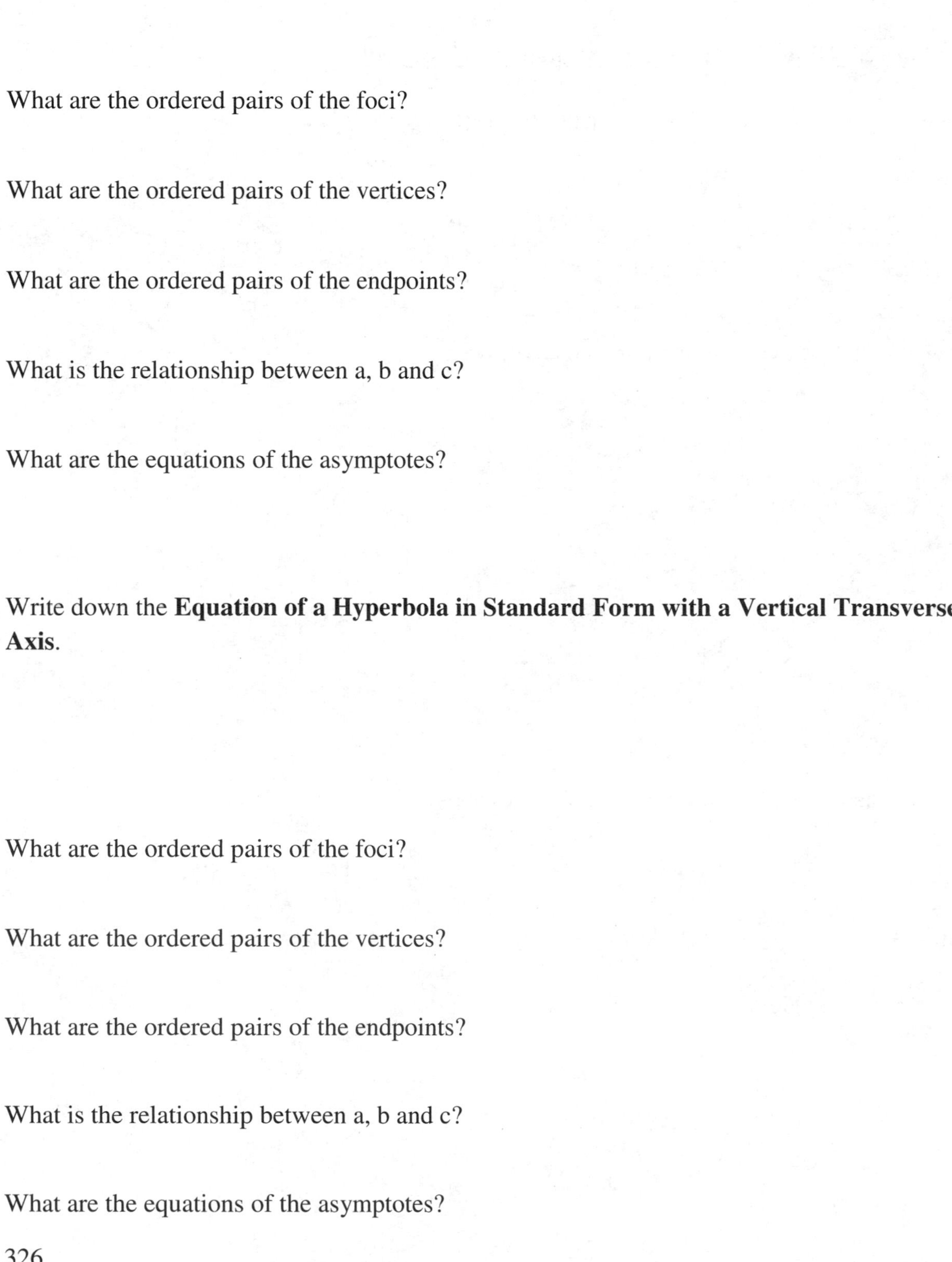

Write down the **Equation of a Hyperbola in Standard Form with a Horizontal Transverse Axis**.

What are the ordered pairs of the foci?

What are the ordered pairs of the vertices?

What are the ordered pairs of the endpoints?

What is the relationship between a, b and c?

What are the equations of the asymptotes?

Write down the **Equation of a Hyperbola in Standard Form with a Vertical Transverse Axis**.

What are the ordered pairs of the foci?

What are the ordered pairs of the vertices?

What are the ordered pairs of the endpoints?

What is the relationship between a, b and c?

What are the equations of the asymptotes?

Write down the **Standard Equations of a Hyperbola with the Center at the Origin.**

What are the ordered pairs of the foci for a *horizontal transverse axis*? For a *vertical transverse axis*?

What are the ordered pairs of the vertices for a *horizontal transverse axis*? For a *vertical transverse axis*?

What are the ordered pairs of the endpoints for a *horizontal transverse axis*? For a *vertical transverse axis*?

What is the relationship between a, b and c?

What are the equations of the asymptotes for a *horizontal transverse axis*? For a *vertical transverse axis*?

Section 12.3

Example 1:
Study the solutions for example 1, showing all steps below. Watch the video for further explanation, if needed.

Sketch the following hyperbolas. Determine the center, transverse axis, vertices, and foci and find the equations of the asymptotes.

a. $\dfrac{(y-4)^2}{36}-\dfrac{(x+5)^2}{9}=1$

$25x^2 - 16y^2 = 400$

Section 12.3 Objective 2: Find the Equation of a Hyperbola in Standard Form

Example 2:
Study the solution for example 2, showing all steps below. Watch the video for further explanation, if needed.

Find the equation of the hyperbola with the center at (-1,0), a focus at (-11,0), and a vertex at (5,0).

Section 12.3 Objective 3: Complete the Square to Find the Equation of a Hyperbola in Standard Form

Example 3:
Study the solution for example 3, showing all steps below. Watch the video for further explanation, if needed.

Find the center, vertices, foci, and equations of asymptotes and sketch the hyperbola

$12x^2 - 4y^2 - 72x - 16y + 140 = 0$

Section 12.3 Objective 4: Solve Applications Involving Hyperbolas

Example 4:
Study the solution for example 4, showing all steps below.

One transmitting station is located 100 miles due east from another transmitting station. Each station simultaneously sends out a radio signal. The signal from the west tower is received by a ship $\frac{1600}{3}$ microseconds after the signal from the east tower. If the radio signal travels at 0.18 miles per microsecond, find the equation of the hyperbola on which the ship is presently located.

Section 13.1 Guided Notebook

Section 13.1 Introduction to Sequences and Series

Read the list of "THINGS TO KNOW" and review any concepts you are unfamiliar with.

Section 13.1 Objective 1: Write the Terms of a Sequence

Write down the definition of a **Finite Sequence** and give one example.

Write down the definition of an **Infinite Sequence** and give one example.

Write down the definition of **The Factorial of a Non-Negative Integer**.

Example 1:
Study the solutions for example 1, showing all steps below. Watch the video for further explanation, if needed.

Write the first four terms of each sequence whose nth term is given.

a. $a_n = 2n - 1$ b. $b_n = n^2 - 1$ c. $c_n = \dfrac{3^n}{(n-1)!}$ d. $d_n = (-1)^n 2^{n-1}$

Section 13.1 Objective 2 Write the Terms of a Recursive Sequence

Write down the definition of a **recursive sequence**.

Example 2:
Study the solutions for example 2, showing all steps below. Watch the video for further explanation, if needed.

Write the first four terms of each of the following recursive sequences.

a. $a_1 = -3, a_n = 5a_{n-1} - 1$ for $n \geq 2$

b. $b_1 = 2$, $b_n = \dfrac{(-1)^{n-1}n}{b_{n-1}}$ for $n \geq 2$

Example 3:
Study the solution for example 3, showing all steps below.

The **Fibonacci sequence** is recursively defined by $a_n = a_{n-1} + a_{n-2}$ where $a_1 = 1$ and $a_2 = 1$. Write the first eight terms of the Fibonacci sequence.

Section 13.1 Objective 3: Write the General Term for a Given Sequence

Example 4:
Study the solutions for example 4, showing all steps below. Watch the video for further explanation, if needed.

Write a formula for the nth term of each infinite sequence then use this formula to find the 8th term of the sequence.

a. $\dfrac{1}{1}, \dfrac{1}{2}, \dfrac{1}{3}, \dfrac{1}{4}, \dfrac{1}{5}, \ldots$

b. $-\dfrac{2}{1}, \dfrac{4}{2}, -\dfrac{8}{6}, \dfrac{16}{24}, -\dfrac{32}{120}, \ldots$

Section 13.1 Objective 4: Compute Partial Sums of a Series

Write down the definition of a **finite series**.

Write down the definition of an **infinite series**.

The sum of the first n terms of a series is called the ______________________ of the series and is denoted as _________.

Example 5:
Study the solutions for example 5, showing all steps below.

Given the general term of each sequence, find the indicated partial term.

a. $a_n = \frac{1}{n}$ find S_3

b. $b_n = (-1)^n 2^{n-1}$ find S_5

Section 13.1 Objective 5: Determine the Sum of a Finite Series Written in Summation Notation

Illustrate **Summation Notation** for both a finite series and an infinite series.

Using one of your examples above, identify the **index of summation**, the **lower limit of summation**, and the **upper limit of summation**.

Example 6:
Study the solutions for example 6, showing all steps below. Watch the video for further explanation, if needed.

Find the sum of each finite series.

a. $\sum_{i=1}^{5} i^2$ b. $\sum_{j=2}^{5} \frac{j-1}{j+1}$ c. $\sum_{k=0}^{6} \frac{1}{k!}$

<u>**Section 13.1 Objective 6**: Write a Series Using Summation Notation</u>

Example 7:
Study the solutions for example 7, showing all steps below.

Rewrite each series using summation notation. Use 1 as the lower limit of summation.

a. $2 + 4 + 6 + 8 + 10 + 12$ b. $1 + 2 + 6 + 24 + 120 + 720 + \ldots + 3{,}628{,}800$

Section 13.2 Guided Notebook

Section 13.2 Arithmetic Sequences and Series

Read the list of "THINGS TO KNOW" and review any concepts you are unfamiliar with.

Section 13.2 Objective 1: Determine if a Sequence is Arithmetic

Write down the definition of an **Arithmetic Sequence** and provide one example.

In an arithmetic sequence the general term has the form $a_n =$ ________________________,
where a_1 is ________________________ and d is ________________________.

Example 1:
Study the solutions for example 1, showing all steps below. Watch the video for further explanation, if needed.

For each of the following sequences, determine if it is arithmetic. If the sequence is arithmetic, find the common difference.

a. $1, 4, 7, 10, 13, \ldots$ b. $b_n = n^2 - n$ c. $a_n = -2n + 7$ d. $a_1 = 14$, $a_n = 3 + a_{n-1}$

When the common difference (d) of an arithmetic sequence is positive, the terms of the sequence________________________ and the graph is represented by a set of ordered pairs that lies along a line with a _______________ slope.

When the common difference (d) of an arithmetic sequence is negative, the terms of the sequence________________________ and the graph is represented by a set of ordered pairs that lies along a line with a _______________ slope.

Section 13.2 Objective 2: Find the General Term or a Specific Term of an Arithmetic Sequence

Rewrite the general term of an arithmetic sequence here: | $a_n =$ |

Example 2:

Study the solutions for example 2, showing all steps below. Watch the video for further explanation, if needed.

Find the general term of each arithmetic sequence, then find the indicated term of the sequence.

a. 11,17,23,29,35,...; a_{50}

b. 2,0,-2,-4,-6,...; a_{90}

c. Find a_{31}

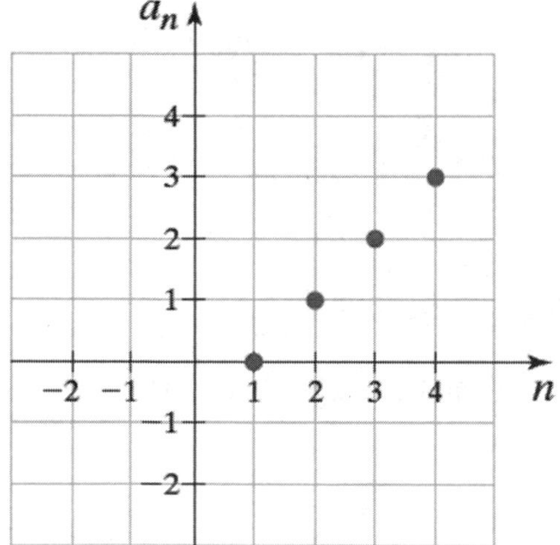

Example 3:
Study the solutions for example 3, showing all steps below. Watch the video for further explanation, if needed.

Find the specific term of an arithmetic sequence:

a. Given an arithmetic sequence with $d = -4$ and $a_3 = 14$, find a_{50}.

b. Given an arithmetic sequence with $a_4 = 12$ and $a_{15} = -10$, find a_{41}.

<u>**Section 13.2 Objective 3:** Compute the *n*th Partial Sum of an Arithmetic Series</u>

Write down the formula for the ***n*th Partial Sum of an Arithmetic Series.** Identify what each variable represents.

Example 4:
Study the solution for example 4 part a, and record the answer below. Complete part b on your own. If your answer is incorrect watch the video to find your error.

Find the sum of each arithmetic series.

a. $\sum_{i=1}^{20}(2i-11)$

b. -5 + (-1) + 3 + 7 + … + 39

<u>**Section 13.2 Objective 4:** Solve Applications of Arithmetic Sequences and Series</u>

Example 5:
Study the solution for example 5.

A local newspaper has hired teenagers to go door-to-door to try to solicit new subscribers. The teenagers receive $2 for selling the first subscription. For each additional subscription sold, the newspaper will pay the teenagers 10 cents more than what was paid for the previous subscription. How much will the teenagers get paid for selling the 100th subscription? How much money will the teenagers earn by selling 100 subscriptions?

Example 6:
Work through example 6 showing all work below. Click on the video to check your work and answer.

A large multiplex movie house has many theaters. The smallest theater has only 12 rows. There are six seats in the first row. Each row has two seats more than the previous row. How many total seats are there in this theater?

Section 13.3 Guided Notebook

Section 13.3 Geometric Sequences and Series

Read the list of "THINGS TO KNOW" and review any concepts you are unfamiliar with.

Section 13.3 Objective 1: Write the Terms of a Geometric Sequence

Write down the definition of a **Geometric Sequence** and provide one example.

In a geometric sequence the general term has the form a_n = ______________________,
where a_1 is __________________________ and r is ____________________________.

Make note of the differences between the graph of an arithmetic sequence and the graph of a geometric sequence (figure 3 on 13.6-5).

Example 1:
Study the solutions for example 1, showing all steps below. Watch the video for further explanation, if needed.

a. Write the first five terms of the geometric sequence having a first term of 2 and a common ratio of 3.

b. Write the first five terms of the geometric sequence such that $a_1 = -4$ and $a_n = -5a_{n-1}$ for $n \geq 2$.

Section 13.3 Objective 2: Determine If a Sequence Is Geometric

How can you determine if a given sequence is geometric?

Example 2:
Study the solutions for example 2 parts a and b, and record the answers below. Complete part c on your own and check your answer by watching the video.

For each of the following sequences, determine if it is geometric. If the sequence is geometric, find the common ratio.

a. $2,4,6,8,10,\ldots$ b. $\frac{2}{3},\frac{4}{9},\frac{8}{27},\frac{16}{81},\frac{32}{243},\ldots$ c $12,-6,3,-\frac{3}{2},\frac{3}{4},\ldots$

__**Section 13.3 Objective 3:** Find the General Term or a Specific Term of a Geometric Sequence__

Rewrite the general term of a geometric sequence here: $a_n =$

Example 3:
Study the solutions for example 3.

Example 4:
Study the solutions for example 4, showing all steps below. Watch the video for further explanation, if needed.

a. Find the seventh term of the geometric sequence whose first term is 2 and whose common ratio is -3.

b. Given a geometric sequence such that $a_6 = 16$ and $a_9 = 2$, find a_{13}.

__**Section 13.3 Objective 4:** Compute the *n*th Partial Sum of a Geometric Series__

(Note the slight difference in the summation notation for the **infinite geometric series** and the summation notation for ***n*th partial sum** of a series).

Write down the formula for the ***n*th Partial Sum of a Geometric Series.** Identify what each variable represents.

Example 5:

Study the solution for example 5 part a, and record the answer below. Complete part b on your own. If your answer is incorrect watch the video to find your error.

a. Find the sum of the series $\sum_{i=1}^{15} 5(-2)^{i-1}$

b. Find the 7th partial sum of the geometric series $8+6+\frac{9}{2}+\frac{27}{8}+....$

Section 13.3 Objective 5: Determine if an Infinite Geometric Series Converges or Diverges

Write down the formula for the **Sum of an Infinite Geometric Series.** Identify what each variable represents. What must the value of r be restricted to?

Given the following infinite geometric series, Table 1 shows the sum of 5, 10, 20 and 40 terms.

$$\frac{1}{2}+\frac{1}{3}+\frac{2}{9}+\frac{4}{27}+\frac{8}{81}+\cdots$$

n	$S_n=\frac{a_1(1-r^n)}{1-r}=\frac{\frac{1}{2}\left(1-\left(\frac{2}{3}\right)^n\right)}{1-\frac{2}{3}}$	$r^n=\left(\frac{2}{3}\right)^n$
5	1.3024691	.1316872
10	1.4739887	.0173415
20	1.4995489	.0003007
40	1.4999999	.0000001

Looking at Table 1, it appears that as n increases, the value of S_n ________________________.

Also note, that as n increases, the value of $r^n=\left(\frac{2}{3}\right)^n$ is getting closer to ________________________.

If $|r|<1$, then the infinite geometric series has a finite sum and is said to _______________.

If $|r|\geq 1$, then the infinite geometric series does not have a finite sum and the series is said to _______________.

Example 6:
Study the solutions for example 6, showing all steps below. Watch the video for further explanation, if needed.

Determine whether each of the following series converges or diverges. If the series converges, find the sum.

a. $\sum_{n=1}^{\infty}\frac{1}{2}\left(\frac{2}{3}\right)^{n-1}$ b. $3-\frac{6}{5}+\frac{12}{25}-\frac{24}{125}+\ldots$ c. $12+18+27+\frac{81}{2}+\frac{243}{4}+\ldots$

<u>**Section 13.3 Objective 6**: Solve Applications of Geometric Sequences and Series</u>

Example 7:
Study the solution for example 7.

Suppose that you have agreed to work for Donald Trump on a particular job for 21 days. Mr. Trump gives you two choices of payment. You can be paid \$100 for the first day and an addition \$50 per day for each subsequent day. Or, you can choose to be paid 1 penny for the first day with your pay doubling each subsequent day. Which method of payment yields the most income?

Example 8:
Study the solution for example 8.

A local charity received \$8,500 in charitable contributions during the month of January. Because of a struggling economy, it is projected that contributions will decline each month to 95% of the previous month's contributions. What are the expected contributions for the month of October? What is the total expected contributions that this charity can expect at the end of the year?

Amount of an Ordinary Annuity after the kth Payment

The total amount of an ordinary annuity after the kth payment is given by the formula

$$A = \frac{P((1+i)^k - 1)}{i}$$

where $A =$ Total amount of annuity after k payments

$P =$ Deposit amount at the end of each payment period

$i =$ Interest rate per payment period

Example 10:
Work through example 10 showing all steps below. Click on the video to check your work and answer.

Chie and Ben decided to save for their newborn son Jack's college education. They decided to invest $200 every 3 months in an investment earning 8% interest compounded quarterly. How much is this investment worth after 18 years?

Section 13.4 Guided Notebook

Section 13.4 The Binomial Theorem

Read the list of "THINGS TO KNOW" and review any concepts you are unfamiliar with.

Section 13.4 Objective 1: Expand Binomials Raised to a Power Using Pascal's Triangle

Watch the video on 13.4-3 and take notes below:

Determine the expansion of $(a+b)^n$. Use the given values of n.

$n=0$: $(a+b)^0 =$ 1

$n=1$: $(a+b)^1 =$ 1a + 1b

$n=2$: $(a+b)^2 =$ $1a^2 + 2ab + 1b^2$

$n=3$: $(a+b)^3 =$

$n=4$: $(a+b)^4 =$

$n=5$: $(a+b)^5 =$

Fill in the missing blanks for Pascal's Triangle:

$n=0$	1
$n=1$	1 1
$n=2$	1 2 1
$n=3$	1 ___ ___ 1
$n=4$	1 4 6 4 1
$n=5$	1 ___ ___ ___ ___ 1

Continue the expansion for $n=6$ and note the final answer in the expansion of $(a+b)^6$:

Section 13.4

Example 1:
Study the solutions for example 1 parts a and b, and record the answers below. Complete part c on your own and watch the video for further explanation, if needed.

Use Pascal's triangle to expand each binomial.

a. $(x+2)^4$

b. $(x-3)^5$

c. $(2x-3y)^3$

Section 13.4 Objective 2: Evaluate Binomial Coefficients

Write down the **Formula for a Binomial Coefficient.**

Example 2:
Study the solutions for example 2, showing all steps below.

Evaluate each of the following binomial coefficients.

a. $\binom{5}{3}$

b. $\binom{4}{1}$

c. $\binom{12}{8}$

How would you use a graphing calculator to evaluate the problems in example 2?

Section 13.4 Objective 3: Expand Binomials Raised to a Power Using the Binomial Theorem

Write down the **Binomial Theorem**.

Example 3:
Study the solutions for example 3, showing all steps below. Watch the video for further explanation, if needed.

Use the Binomial Theorem to expand each binomial.

a. $(x-1)^8$

b. $(\sqrt{x}+y^2)^5$

Section 13.4 Objective 4: Find a Particular Term or a Particular Coefficient of a Binomial Expansion

Write down the **Formula for the $(r+1)^{st}$ Term of a Binomial Expansion.**

Example 4:
Study the solution for example 4, showing all steps below. Watch the video for further explanation, if needed.

Find the third term of the expansion of $(2x-3)^{10}$.

Example 5:
Study the solution for example 5, showing all steps below. Watch the video for further explanation, if needed.

Find the coefficient of x^7 in the expansion of $(x+4)^{11}$.

Appendix A Guided Notebook

Appendix A Synthetic Division

Read the list of "THINGS TO KNOW" and review any concepts you are unfamiliar with.

Appendix A Objective 1: Divide a Polynomial by a Binomial Using Synthetic Division

Watch the animation on A-3.

Example 1:
Study the solution for example 1 and record the answer below.

Divide $2x^4+9x^3-12x+1$ by $x+5$ using synthetic division.

Example 2:
Work through example 2 showing all steps below. Click on the link to check your answer. If your answer is incorrect, watch the video to find your error.

Divide $2x^4-3x^2+5x-30$ by $x-2$ using synthetic division.

Read and summarize the CAUTION statement on A-7.

Appendix B Guided Notebook

Appendix B Mean, Median, Mode

Read the list of "THINGS TO KNOW" and review any concepts you are unfamiliar with.

Appendix B Objective 1: Find the Mean

Why is it helpful to find a **measure of central tendency** when we are given a set of data values?

In the introduction to this appendix we read that a student's grade point average is an example of a measure of central tendency. Please list two other examples below.

1.

2.

What are the three most common measures of central tendency?

1.

2.

3.

What is another name for the **mean**?

How is the mean computed?

Example 1:
Study the solution for example 1on page B-3, showing all work below. Watch the video for further explanation, if needed.

Valerie keeps track of her gasoline purchases to help her budget monthly expenses. The following data shows the price per gallon (in dollars) for regular unleaded gasoline on her last 10 fill-ups.

2.97	3.00	3.08	3.10	3.04
2.96	3.06	3.07	2.95	3.07

a. Find the mean.

b. How many values are below the mean? How many are above the mean?

c. Suppose the value 3.10 was mistakenly recorded as 31.0. Find the mean using this value.

d. How did changing the value in (c) affect the value of the mean?

<u>**Appendix B Objective 2:** Find the Median</u>

How do you find the **median** if the number of data values in the ordered list is odd?

How do you find the **median** if the number of data values in the ordered list is even?

Example 2:
Study the solution for example 2 on page B-5, showing all work below. Watch the video for further explanation, if needed.

The following data show the number of apps downloaded on 8 randomly selected iPhones.

48 39 43 34 27 22 37 46

a. Find the median.

b. How many values are below the median? How many are above the median?

c. Suppose the value 48 was mistakenly recorded as 84. Find the median using this value.

d. How did changing the value in (c) affect the value of the median?

Read and summarize the CAUTION statement on page B-7.

<u>**Appendix B Objective 3:** Find the Mode</u>

Given a set of data values, how do you find the **mode**?

When would a data set have no **mode**?

When would a data set have one **mode**?

When would a data set have more than one **mode**?

The **mode** is not greatly affected by a few extreme values. What other measure of central tendency is this similar to?

Example 3:
Study the solution for example 3 on page B-8, showing all work below. Watch the video for further explanation, if needed.

The following data are the snowfall readings (in inches) at Boston's Logan International Airport for Boston's top twelve winter storms. (Source: cbsboston.com)

27.5	27.1	26.3	25.4	22.5	21.4
19.8	19.4	18.7	18.2	18.2	18.2

a. Find the mode.

b. Suppose the value 27.5 was mistakenly recorded as 72.5. Find the mode using this value.

c. How did changing the value in (c) affect the value of the mode?

Appendix C Guided Notebook

Appendix C Solving Systems of Linear Equations Using Matrices

Read the list of "THINGS TO KNOW" and review any concepts you are unfamiliar with.

Appendix C Objective 1: Write an Augmented Matrix

What is a matrix? Please give an example.

The size of a matrix is determined by the number of __________ and __________. A 2×3 (read *two by three*) matrix has _____ rows and _____ columns.

What is an augmented matrix?

Write the augmented matrix for the following system of equations:

$$\begin{cases} 3x+2y=11 \\ 8x-9y=4 \end{cases}$$

Example 1:
Study the solutions for parts a and b and record the answers below. Complete part c on your own and check your answer by clicking on the link. If your answer is incorrect, watch the video to find your error.

Write the corresponding **augmented matrix** for each system of equations.

a. $\begin{cases} -2x+7y=9 \\ 8x+3y=0 \end{cases}$

b. $\begin{cases} 6x-3y=11 \\ y=2x-5 \end{cases}$

c. $\begin{cases} 5x-2y+z=18 \\ x+3z=-5 \\ 3y=9+z \end{cases}$

Appendix C Objective 2: Solve Systems of Two Equations Using Matrices

When are two matrices **equivalent matrices**?

Please list the **Row Operations**:

1.
2.
3.

Fill in the table below:

Notation	Meaning
$R_i \Leftrightarrow R_j$	
$kR_i \rightarrow \text{New } R_i$	
$kR_i + R_j \rightarrow \text{New } R_j$	

Watch the **popup** on page C-8 to see an example of each row operation.

Describe a matrix in **row-echelon** form.

Example 2:
Study the solutions for part a and record the answer below. Complete part b on your own and check your answer by clicking on the link. If your answer is incorrect, watch the video to find your error.

Use **matrices** to solve each of the following systems.

a. $\begin{cases} 2x - 5y = 26 \\ 3x + 2y = 1 \end{cases}$ b. $\begin{cases} y = 3x + 2 \\ 6x + \frac{1}{2}y = 6 \end{cases}$

Study the solution to **Example 3** and watch the video.

Appendix C Objective 3: Solve Systems of Three Equations Using Matrices

Writing Row-Echelon Form (Three Rows)

Step 1. ____

Step 2. ____

Step 3. ____

Step 4. ____

Step 5. ____

Example 4:

Study the solutions for part a and record the answer below. Complete part b on your own showing your work below and check your answer by clicking on the link. If your answer is incorrect, watch the video to find your error.

Use **matrices** to solve each of the following systems.

a. $\begin{cases} 2x+y-3z=-3 \\ x+2y+z=4 \\ -3x+y-4z=1 \end{cases}$

b. $\begin{cases} 2x+2y+z=1 \\ 5x+2y=30+3z \\ 3x+4z=-11 \end{cases}$

Example 5:

Work through Example 5 on your own showing your work below and check your answer. If your answer is incorrect, watch the video to find your error.

Use **matrices** to solve each of the following system.

$$\begin{cases} x+2y-3z=5 \\ -2x-3y+4z=-6 \\ 2x+4y-6z=10 \end{cases}$$

Summarize the caution statement on page C-24.

Appendix D Guided Notebook

Appendix D Determinants and Cramer's Rule

Read the list of "THINGS TO KNOW" and review any concepts you are unfamiliar with.

Appendix D Objective 1: Evaluate a 2 x 2 Determinant

A **square matrix** has an equal number of ______________ and ______________.

Please give an example of a **square matrix**.

Write the definition of the Determinant of a 2×2 Matrix.

Read the caution statement on page D-4.

Example 1:
Study the solutions for parts a and b and record the answers below.

Evaluate each determinant.

a. $\begin{vmatrix} 5 & 2 \\ 8 & 7 \end{vmatrix}$

b. $\begin{vmatrix} -9 & 3 \\ -4 & 2 \end{vmatrix}$

Appendix D Objective 2: Use Cramer's Rule to Solve a System of Linear Equations in Two Variables

Please write **Cramer's Rule for Solving Systems of Linear Equations in Two Variables**.

Watch the video on page D-6 to see Cramer's Rule derived.

Example 2:
Study the solution for part a and record the answer below. Complete part b on your own and check your answer by clicking on the link. If your answer is incorrect, watch the video to find your error.

Use **Cramer's Rule** to solve each **system**.

a. $\begin{cases} 3x+2y=12 \\ 4x-5y=-7 \end{cases}$ b. $\begin{cases} 6x+y=-2 \\ 9x-2y=11 \end{cases}$

If $D=0$, the $x=\dfrac{D_x}{D}$ and $y=\dfrac{D_y}{D}$ are not ____________________, so **Cramer's Rule** does not apply. However, $D=0$ tells us that the **system** is either _______________ or _______________.

Read through **Cramer's Rule with Inconsistent and Dependent Systems**.

Example 3:
Study the solution for part a and record the answer below. Complete part b on your own and check your answer by clicking on the link. If your answer is incorrect, watch the video to find your error.

Use **Cramer's Rule** to solve each **system**.

a. $\begin{cases} 2x-5y=7 \\ -4x+10y=11 \end{cases}$ b. $\begin{cases} 6x+10y=-50 \\ -9x-15y=75 \end{cases}$

Appendix D Objective 3: Evaluate a 3 x 3 Determinant

For every entry of a 3×3 determinant, there is a 2×2 determinant associated with it called its __________.

For the following matrix, please find the minor associated with b_1.

$$\begin{vmatrix} a_1 & b_1 & c_1 \\ a_2 & b_2 & c_2 \\ a_3 & b_3 & c_3 \end{vmatrix}$$

> **Definition Determinant of a 3 x 3 Matrix**
>
> The determinant of a 3×3 matrix $\begin{bmatrix} a_1 & b_1 & c_1 \\ a_2 & b_2 & c_2 \\ a_3 & b_3 & c_3 \end{bmatrix}$ is denoted by $\begin{vmatrix} a_1 & b_1 & c_1 \\ a_2 & b_2 & c_2 \\ a_3 & b_3 & c_3 \end{vmatrix}$ and is defined by
>
> $$\begin{vmatrix} a_1 & b_1 & c_1 \\ a_2 & b_2 & c_2 \\ a_3 & b_3 & c_3 \end{vmatrix} = a_1\begin{vmatrix} b_2 & c_2 \\ b_3 & c_3 \end{vmatrix} - b_1\begin{vmatrix} a_2 & c_2 \\ a_3 & c_3 \end{vmatrix} + c_1\begin{vmatrix} a_2 & b_2 \\ a_3 & b_3 \end{vmatrix}.$$

Please study the solution for **Example 4**.

We can find the value of a determinant by expanding the minors across any row or down any column. However, we must be careful to properly add or subtract the product of each row or column entry and its minor.

If the **sum** of the row number and the column number is __________, then the **product** of the entry and its minor is "added." If the sum of the row number and column number of an entry is __________, then the product is "subtracted."

Example 5:

Study the solution for part a and record the answer below. Complete parts b and c on your own. Watch the interactive video for the complete solutions.

Evaluate the **determinant** by expanding the minors for the given row or column.

a. First column

Example 5 (continued)

b. Second row

c. Third column

<u>**Appendix D Objective 4:** Use Cramer's Rule to Solve a System of Linear Equations in Three Variables</u>

Cramer's Rule for Solving Systems of Linear Equations in Three Variables

For the system of linear equations $\begin{cases} a_1x + b_1y + c_1z = n_1 \\ a_2x + b_2y + c_2z = n_2 \\ a_3x + b_3y + c_3z = n_3 \end{cases}$, the solution is given by

$$x = \frac{D_x}{D}, \quad y = \frac{D_y}{D} \quad \text{and} \quad z = \frac{D_z}{D}$$

where

$$D = \begin{vmatrix} a_1 & b_1 & c_1 \\ a_2 & b_2 & c_2 \\ a_3 & b_3 & c_3 \end{vmatrix}, \quad D_x = \begin{vmatrix} n_1 & b_1 & c_1 \\ n_2 & b_2 & c_2 \\ n_3 & b_3 & c_3 \end{vmatrix}, \quad D_y = \begin{vmatrix} a_1 & n_1 & c_1 \\ a_2 & n_2 & c_2 \\ a_3 & n_3 & c_3 \end{vmatrix}, \quad \text{and } D_z = \begin{vmatrix} a_1 & b_1 & n_1 \\ a_2 & b_2 & n_2 \\ a_3 & b_3 & n_3 \end{vmatrix},$$

provided $D \neq 0$.

Example 6:

Study the solution for part a and record the answer below. Complete part b on your own and check your answer by clicking on the link. If your answer is incorrect, watch the video to find your error.

Use **Cramer's Rule** to solve each **system**.

a. $\begin{cases} x-y+4z=-5 \\ 6x-27+5z=1 \\ 3x+2y+2z=10 \end{cases}$

b. $\begin{cases} x+y+z=3 \\ 2x-3y=14 \\ 4y+5z=-3 \end{cases}$

Read through Cramer's Rule with Inconsistent and Dependent Systems in Three Variables.